T0137255

Springer Topics in Signal Processing

Volume 11

Series editors

Jacob Benesty, Montreal, Canada
Walter Kellermann, Erlangen, Germany

More information about this series at http://www.springer.com/series/8109

Mohammad Ali Nematollahi · Chalee Vorakulpipat
Hamurabi Gamboa Rosales

Digital Watermarking

Techniques and Trends

 Springer

Mohammad Ali Nematollahi
National Electronics and Computer
   Technology Center (NECTEC)
Pathumthani
Thailand

Chalee Vorakulpipat
National Electronics and Computer
   Technology Center (NECTEC)
Pathumthani
Thailand

Hamurabi Gamboa Rosales
Universidad Autónoma de Zacatecas
Zacatecas
Mexico

ISSN 1866-2609   ISSN 1866-2617   (electronic)
Springer Topics in Signal Processing
ISBN 978-981-10-9527-6   ISBN 978-981-10-2095-7   (eBook)
DOI 10.1007/978-981-10-2095-7

Printed on acid-free paper

This Springer imprint is published by Springer Nature
The registered company is Springer Science+Business Media Singapore Pte Ltd.

To my father Abo-Al-Ghasem, my mother Ziba, my sisters Nazanin, Mahnaz, Saeedeh, Freshteh, and my nieces Yekta & Anahita.

Preface

Due to wide range of digital watermarking applications in digital world, available digital watermarking books just concentrate on specific application of digital watermarking such as multimedia, document, and network. This book is written to cover all possible application of digital watermarking in digital world. For this purpose, the overall techniques and trends in each of the digital watermarking forms are discussed with simple and understandable language. Since digital watermarking has broad contribution in cybersecurity science, the authors believe that any sort of digital watermarking for digital contents requires a separate chapter for complete discussion. Therefore, every chapter of this book is dedicated to well-known application of digital watermarking in audio, speech, image, video, 3D graphic, natural language, text, software, XML data, relational database, ontology, network stream, DNA, Bitcoin, and hardware IPs.

This book is suitable for beginners who do not have any information about digital watermarking concepts. In order to motivate readers to select, familiar, and get some concepts of a desired watermarking field, this book is structured and organized in such a way that the reader has ability to only focus on a chapter. For more and deep concentration on details for that chapter, the reader must study state-of-the-art references in that chapter. For course work usage, the authors suggest students to familiar with basic concepts of digital signal processing, natural langue processing, software engineering, network engineering, machine learning, digital design, and cryptography subjects.

Although the authors have been attempted to provide the best suggestion for classification of techniques and trends in the beginning of each chapter, the authors declare that there is always improvement for classifying these techniques and trends. Therefore, any suggestion from audience for enhancing the book's structure and book's content will be most welcomed.

Pathumthani, Thailand   Mohammad Ali Nematollahi
Pathumthani, Thailand   Chalee Vorakulpipat
Zacatecas, Mexico   Hamurabi Gamboa Rosales

Acknowledgements

First and foremost, I would like to thank and pray to God for blessing my research study and giving me strength and courage.

I also would like to thank my father and mother for their encouragement and support by praying for me to do my book successfully.

Furthermore, I would like to express my gratitude and convey my thanks to Dr. Hamurabi Gamboa Rosales and Dr. Chalee Vorakulpipat for their supervision, advice, and guidance from the early stage of this book until the completion of the book.

Similarly, I would like to record my gratitude to my Editors Dr. Loyola D'Silva and Mr. Ravi Vengadachalam for their guidance, advice, and continuous support during the entire course of this book.

Lastly, I would like to express my special thanks to Thailand's National Electronics and Computer Technology Center (NECTEC) for providing a beautiful, peaceful, and calm academic environment for research and study.

Contents

About the Authors

Mohammad Ali Nematollahi was born in 1986 in Shiraz, Iran. He received his Bachelor of Science degree in Computer Engineering (software engineering) in Yazd, Iran, in 2008. He completed his Master's degree in Computer Engineering (software engineering) in Islamic Azad University (IAU), Dubai, UAE, in 2011. He also was a lecturer in IAU in 2010 and 2011. He is a Ph.D graduate in Computer and Embedded System Engineering from Universiti Putra Malaysia (UPM) in 2015. He is currently a postdoctorate researcher in Cybersecurity Laboratory at the National Electronics and Computer Technology Center (NECTEC) of Thailand. He had written numerous articles which were mainly published in International Journals. His research interests include digital signal processing, speaker recognition, and digital watermarking.

Contact him: greencomputinguae@gmail.com

Dr. Chalee Vorakulpipat received his B.Eng. in electronics engineering from King Mongkut's Institute of Technology Ladkrabang, Thailand, and M.S. in information technology from Kasetsart University, Thailand. He was awarded a scholarship from the Royal Thai Government to pursue a doctoral study. He earned his Ph.D. in information systems from the University of Salford, UK. He is currently a senior researcher and head of Cybersecurity Laboratory at the National Electronics and Computer Technology Center of Thailand. He has been involved in several projects in information security, mobile device management, social networking sites, ubiquitous computing, context-aware computing, e-health, and mobile application development. He has over 30-refereed publications in these areas which appeared in conference proceedings and journals such as Computers and Security, Advanced Engineering Informatics, Automation in Construction, Knowledge Engineering Review, and ETRI Journal. He also serves as subcommittee on national information security of Thailand. In the academic role, he works as a lecturer for information systems courses at several universities in Thailand. He holds information security professional certificates including CISSP, CISA, and IRCA: ISMS Lead Auditor, and a project management professional certificate—PMP.

Contact him: Chalee.Vorakulpipat@nectec.or.th

Hamurabi Gamboa Rosales received his Bachelor in Electronics and Communications Engineering in the Faculty of Engineering of the University of Guadalajara in 2000. From 2001 to 2003, He completed his Master's degree in Electrical Engineering from, focusing on the Digital Signal Processing at the University of Guanajuato. He completed his doctoral studies at the Technical University of Dresden, Germany, in the area of Voice Processing 2010. Currently he is working as a professor and researcher at the Academic Unit of Electrical Engineering of the Autonomous University of Zacatecas, Mexico, in the area of research digital signal processing.

Contact him: hamurabigr@uaz.edu.mx

Abbreviations

AbS	Analysis-by-synthesis
AD	Analog to digital
ADPCM	Adaptive differential pulse code modulation
APF	All pass digital filters
AR	Autoregressive
ATC	Air traffic control
AWGN	Additive white Gaussian noise
BER	Bit error rate
BPF	Band-pass filter
BR	Block replacement
CELP	Code excitation linear prediction
DA	Digital to analog
DBA	Database base administrator
DCT	Discrete cosine transform
DEW	Differential energy watermark
DFT	Discrete fourier transform
DNA	Deoxy-ribonucleic Acid
DRT	Diagnostic rhyme test
DSP	Digital signal processing
DSSS	Direct sequence spread spectrum
DWT	Discrete Wavelet Transform
ELT	Extended lapped transforms
FFT	Fast fourier transform
FNR	False negative rate
FPGA	Field programmable gate array
FPR	False positive rate
FSM	Finite state machine
HAS	Human auditory system
HDL	Hardware design language
HVS	Human visual system

IBW	Interval-based watermarking
ICA	Independent component analysis
ICBW	Interval centroid-based watermarking
IP	Intellectual Property
IPD	Inter-packet delays
IPR	Intellectual Property Rights
JAWS	Just another watermarking system
LAR	Log area ratio
LM	Language model
LPA	Linear predictive analysis
LPC	Linear predictive coefficient
LPF	Low-pass filter
LSB	Least significant bit
LSP	Line spectrum pair
LUT	Lookup table
LWT	Lifted wavelet transform
MFA	Multi-flow attack
MFA	Multi-factor authentication
MFCC	Mel frequency cepstrum coefficients
M-JPEG	Moving Joint Photographic Experts Group
ML	Maximum-likelihood
MLT	Modulated lapped transforms
MNB	Measuring normalized blocks
MOL	Maximum occurring letter
MOS	Mean opinion score
MPA	Modified patchwork algorithm
MPEG	Motion Picture Experts Group
MRMS	Maximum root mean square
MRT	Modified rhyme test
MSE	Mean-squared error
MST	Minimum spanning tree
NLG	Natural language generation
NLP	Natural language processing
NURBS	Non-uniform rational basis spline
OCR	Optical character recognition
OLPP	Orthogonal locality preserving projections
OS	Operating system
PCA	Principle component analysis
PCFG	Probabilistic context-free grammar
PESQ	Perceptual evaluation of speech quality
PMR	Pattern matching rate
PN	Pseudo-noise
PSNR	Peak signal-to-noise ratio
PSQM	Perceptual Speech Quality Measure

QEM	Quadric error metric
QIM	Quantization index modulation
QP	Quadratic programming
QPS	Quantitative perfusion SPECT
RBA	Random bending attack
RDM	Rational dither modulation
RDWT	Redundant discrete wavelet transform
RMSE	Root mean-squared error
RNG	Random number generator
RTL	Register transfer level
SAT	constraint-SATisfaction
SD	Spectral distortion
SDG	Subjective difference grade
SEGSNR	Segmental signal-to-noise ratio
SNR	Signal-to-noise ratio
SOC	System-on-a-chip
SS	Spread spectrum
STG	State transition graph
SVD	Singular value decomposition
TFA	Temporal frame averaging
TMR	Text meaning representation
VEP	Vector extraction paradigm
VHF	Very high frequency
VLC	Variable length coded
VoIP	Voice over IP
VSI	Virtual socket interface
WDR	Watermark distortion rate
WECR	Watermark estimation clustering and remodulation
WER	Watermark estimation remodulation
WSED	Watermarking subspace estimation draining
XML	Extensible markup language
ZCR	Zero crossing rate

List of Figures

List of Tables

Chapter 1
Preliminary on Watermarking Technology

1.1 Overview

A recent development in communication technology, storage device, and digital recording has created an Internet environment with the ability to distribute, duplicate, obtain, and replicate digital media without any quality lossless. However, this technology advancement is worrying multimedia content publishing industries due to unauthorized access through digital media content which is requiring a serious and immediate protection for Intellectual Property Rights. Although a traditional cryptography approach can apply to prevent an unauthorized party to access a digital media content by encrypting the content of the media, cryptography technology has some limitations to fully protect the Intellectual Property Rights. Therefore, it is obvious that other technology should be applied to construct the Intellectual Property Rights, track the digital media content, and provide the digital media content authentication which ensure authorized access and prevent illegal manipulation to the content of digital media. As an effective solution, **Digital Watermarking** is applying to facilitate the requirements for protecting the Intellectual Property Rights of a digital media. Currently, digital watermarking is becoming the hot topic for the research communities that actively working on this industry. The idea of digital watermarking is the hiding information about digital content as a metadata within digital content to protect the ownership. Probably, this idea is invented in Germany due to a German term "wassermark" is available that means effect of water on paper [1].

Apart from the basic concepts in data hiding, this book is attempting to cover recent and practical applications of digital watermarking in all computer science fields in order to provide an overview on different roles of digital watermarking in computer science. Figure 1.1 shows the main branch in data-hiding technology. It presents the main focus of this book to narrow down the constraints in the field of

© Springer Science+Business Media Singapore 2017
M.A. Nematollahi et al., *Digital Watermarking*, Springer Topics
in Signal Processing 11, DOI 10.1007/978-981-10-2095-7_1

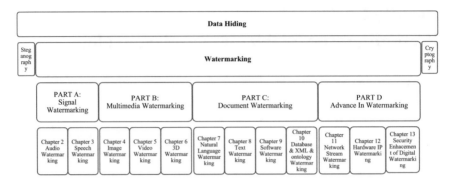

Fig. 1.1 Book focus in the field of data hiding

data hiding. As shown, each chapter of this book is dedicated to an application of digital watermarking in order to provide a deep and comprehensive overview on major digital watermarking technology.

1.2 Fundamentals of Digital Watermarking

The art of watermarking is hiding extra information (which could be an image logo, text message, and raw watermark bits) inside the content of the host objects such as images, audio signals, speech signals, 3D graphical objects, videos, texts, software codes, network streams, XML data, and ontologies without serious degradation on the quality of the objects. The watermark must be detectable from the watermarked content even when intentional and unintentional manipulations have been conducted on the digital watermarked content. Therefore, digital watermarking can be expressed by two main processes: watermark embedding and watermark extraction, which are illustrated in Fig. 1.2.

The watermarking processes can be described formally. For creating a watermarked cover W_C, a watermark data W should be embedded into a cover object O_c. As a result, watermark distortion is defined by the deference between O_c and W_C. Usually, the raw watermark data W is encoded by using a secret key K in order to

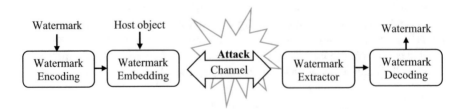

Fig. 1.2 Concept of digital watermarking

improve the security and decrease the watermark payload P. Then, a modulated/scaled watermark method W_M is used to embed watermark bits into the cover object O_c with minimum embedding distortion to provide enough imperceptibility. After embedding the watermark, the watermarked cover W_C may be imposed to intentional and unintentional manipulations such as conversions, compression, noise, adding, removing, and editing which would change it to manipulated watermarked cover $\widehat{W_C}$. The amount of degradation can be computed by finding the deference between $\widehat{W_C}$ and W_C which may be considered as noise of the environment.

Detection of the watermark is done based on the received manipulated watermarked cover $\widehat{W_C}$ and the watermark key K. There are two major approaches for watermark detection: The first one is informed detector (non-blind) approach that requires original cover work O_c, and the second one is oblivious detector (blind) approach which extracts watermark only from manipulated watermarked cover $\widehat{W_C}$ without any knowledge of the original cover O_c. Thus, digital watermarking can be categorized based on source and extraction module into three main classes:

(a) Blind watermarking that detects the watermark without the original cover O_c. However, in some applications, key K uses are required for generating random sequences.
(b) Semi-blind watermarking that detects the watermark with using some information about the original cover O_c. For example, it should access to the published watermarked signal that is the original signal after just adding the watermarks.
(c) Non-blind watermarking that detects the watermark using the original cover O_c and manipulated watermarked cover $\widehat{W_C}$ together.

There is always possibility to categorize the watermarking techniques based on various factors. With respect to robustness, the watermarking methods can be divided into three main groups:

(a) Robust digital watermarking that detects the watermark even under serious and malicious manipulation.
(b) Semi-fragile digital watermarking that detects the watermark just under unintentional manipulation.
(c) Fragile digital watermarking that detects the watermark only if no intentional and unintentional manipulations are took place. This type of watermark is needed for paper of bank notes. These watermarks do not survive any kind of copying and can be used to indicate the bill's authenticity. Reaching for fragility is more difficult than robustness.

1.3 Requirements of Digital Watermarking

Basically digital watermarking can be considered as a tradeoff problem due to the nature of digital watermarking which is competing among payload, transparency, robustness, and security. Depending on various and wide applications of the digital

Fig. 1.3 Requirements of
digital watermarking [2]

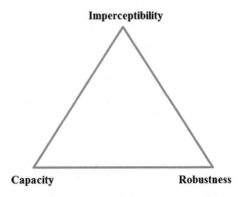

watermarking, the watermarking system is designed based on these requirement and system inherent properties. Figure 1.3 illustrates the requirements of digital watermarking. As shown, these requirements oppose one another and to make them meet is difficult. Within this magic triangle, "the mode of operation" of a particular watermarking system can be set. For example, a high data rate system may enhance its rate by lowering the robustness at the same moment.

There are other properties such as fidelity, tamper resistance, false positive rate, and computational cost which may be required in some applications. In real application, designing a prefect watermarking system that facilitates all watermarking requirement is impossible or if to be positive, it is much more difficult task. Therefore, giving priority to each watermark requirement is necessary which must be performed with special attention and careful analysis of the application. In addition, the major requirements of digital watermarking are discussing to clear terminology definition with various application.

1.3.1 Robustness of Digital Watermarking

While the watermarked object is published and distributed, it is subjected to common operations such as conversion, compression, addition, editing, and removing which may influence on watermark extraction process. Although the watermark must be extractable under these situations, the extracted watermark is not necessary similar to original one. Whatever, the extracted watermark has less difference with original watermark under intentional and unintentional attack, digital watermarking is tending to be robust. Considering robustness as a single-dimensional value is incorrect because each watermarking technique is weak against at least an attack. Moreover, designing a watermarking technique which is robust to all types of manipulation is unnecessary and excessive. For example, in radio and television broadcast monitoring application, there are needs that watermark is robust against transmission modification including A/D, D/A conversions, compressions, vertical, and horizontal translations but there is no need that watermark is robust against wide variety of modifications such as scaling, rotation, filtering, and distortion that might not take place during broadcasting.

In addition, other types of applications require other types of covert communication including steganography and fragility in which robustness is totally undesirable and irrelevant for that application. For example, authentication only requires fragile watermarking to check whether the digital object has been modified. To conclude, robustness requires just for unpredictable applications such as proof of ownership, copy control, identification, and fingerprinting where conceivable distortion is inevitable and there is a concern about removal of the watermark.

1.3.2 Capacity of Digital Watermarking

The hidden bit rate in digital object is defined as capacity or payload and usually is measured in bit per second (bps). In some cases, the watermark bit is inherently tied to the number of alternative messages that can be embedded thanks to the encoding algorithm. It is obvious that the payload is directly related to the size of the host data. Whenever the amount of the host samples increases, more watermark bits embed into the host data.

1.3.3 Imperceptibility of Digital Watermarking

Imperceptibility is defined as the amount of distortion which is injected by embedding the watermark. Imperceptibility can be expressed by measuring quality or fidelity. In contrast to fidelity that measures the similarity between original and watermarked objects, an independent acceptability technique is applied on watermarked object to measure quality of the watermarked technique. There are two tests available for fidelity and quality measurements including subjective and objective measurements. The most reliable and accurate test for both quality and fidelity is the subjective test that involves human observation and psychophysics model. Subjective test is conducted based on a scientific discipline in order to determine the relationship between the physical world and the people's subjective experience of that world.

1.3.3.1 Objective Measurement

In the objective test, the main aim is expressing the amount of the watermark distortion which can be computed by comparing similarity or deference between original and watermarked objects. A simple and common method known as signal-to-noise ratio (SNR) can be a good example of objective measurement which tries to quantify the watermarked distortion by expressing it as a noise. The objective tests are varying for different digital contents. In the case of audio and speech watermarking, the main tests include spectral distortion (SD), root-mean-squared error (RMSE), mean squared error (MSE), peak signal-to-noise ratio (PSNR), and segmental signal-to-noise ratio (SEGSNR).

1.3.3.2 Subjective Measurement

There is not always correlation between objective measurement and human perceptions. On the other hands, human response to a watermarked object may be different from the result of objective measurements. Therefore, subjective measurement is used to better evaluate watermarking techniques. The most common subjective technique is mean opinion score (MOS) which is more simple and popular than other subjective techniques. It is based on reporting the dissimilarities between the quality of the original and watermarked objects. For instance, Table 1.1 presents MOS grades for speech watermarking techniques based on the understanding of the speech signals.

Subjective listening tests are conducted by human's auditory and human's visually perceptions. It should be mentioned that Human Auditory System (HAS) is much more sensitive than Human Visual System (HVS). As a result, reaching inaudibility for audio and speech signals is much more difficult than invisibility in which both are related to imperceptibility requirement of digital watermarking.

1.3.3.3 Intelligibility Measurement

Intelligibility means how understandable the digital object is. The main focus of intelligibility is on information value and content of digital media. For example, high-quality voice of a baby speech has lack of verbal information which causes no intelligibility at all although the sound quality is pleasant. For majority of applications such as speech, text, and natural language, intelligibility is much more desirable than imperceptibility [4]. Here some intangibility measurement techniques for speech include the following: Measuring Normalized Blocks (MNB), Modified Rhyme Test (MRT), Perceptual Speech Quality Measure (PSQM), Perceptual Evaluation Of Speech Quality (PESQ), Subjective Difference

Table 1.1 MOS grades [3]

MOS	Quality	Quality scale	Effort to understand the meaning scale
5	Excellent	Imperceptible	No effort required
4	Good	Perceptible, but not annoying	No appreciable effort required
3	Fair	Slightly annoying	Moderate effort required
2	Poor	Annoying	Considerable effort required
1	Bad	Very annoying	No meaning was understood

Grade (SDG), Diagnostic Rhyme Test (DRT), phonetically balanced word lists, diagnostic alliteration test, diagnostic medial consonant test, ICAO spelling alphabet test, two-alternative forced choice, six-alternative rhyme test, general word test, general sentence test, consonant-vowel-consonant test, and four alternative auditory feature test [1].

1.3.4 Tamper Resistance

The resistance of the watermark system against hostile attacks is known as tamper resistance. Although there are several kinds of tamper resistance which are application dependent, certain kinds of these hostile attacks are more important for tamper resistance. In the following, these types of attacks are categorized in four types:

Active attacks: The main aim of this attack is making the watermark undetectable which may be done by removing the watermark. Although active attack is not an issue for covert communication and authentication, it would be a critical issue for some watermark applications such as proof of ownership, owner identification, copy control, and fingerprinting.

Passive attacks: The main aim of this attack is identifying the availability of the watermark which determines whether a mark is present in a covert communication. Although passive attack is not concerned in many watermark application scenarios, it is primary interest for some steganography applications like convert communication to prevent the watermark from being observed.

Collusion attacks: The main aim of this attack is constructing a version of object without watermark by copying several pieces of the object. Collusion attack can be regarded as a special type of active attack. This attack is mainly applied on fingerprinting watermark application which needs several copies of the watermarked object for obtaining any watermark that seems unlikely prospect.

Forgery attacks: The main aim of this attack is trying to fabricate the authentication process by embedding valid watermark into the watermarked object. Forgery attack is a serious concern for some watermarking applications such as proof of ownership to prevent bogus acceptance and modification of the watermarked object.

1.3.5 Computational Cost

Generally, computational cost is referred to dollar cost and speed requirements for both watermark embedding and extraction processes. In the computational cost requirement, the major goal is designing a watermark device with more numerous applications with minimum computational cost. Depending on the watermark application whether user willing to wait or not, the watermark system is designed with different speeds for embedding and extraction. For instance, in broadcast and monitoring applications, the embedding and extraction processes should be worked in a real-time manner. Also, the broadcast application requires a few watermark embedding modulates with several hundred extraction modulates. However, in proof of ownership application, the watermark detection could be consumed much more time like several days to detect the watermark. For copy control application, although only a watermark embedding module is enough a lots of watermark detector modules are required.

1.3.6 False Positive Rate

False positive rate (FPR) is referring to detection of a watermark when there is no actual watermark in the watermarked object. Therefore, it is expected to have FPR during several attempts for watermark extraction from objects. On the other word, the probability of FPR can be computed for a given detector during running. Basically, there are two FPRs which can be computed. In the first FPR, probability computation is fully depended on method of watermark generation, which can be a Gaussian, independent random number, or bit-encoding algorithm. The computation of this FPR is independent from the piece of watermarked object. However, in the second FPR probability computation is independent from watermark and it is dependent on the selected piece of object randomly. Although finding the first FPR has narrow applications like fingerprinting which leads to false accusation of theft, more attentions are attracted to compute the second FPR. The amount of FPR is depended on application which always should be infinitesimal for many water-marked applications as it prevents serious trouble in copy control application.

1.3.7 False Negative Rate

False negative rate (FNR) or false miss is referring to lack of the ability for detect-ing the valid watermark from watermarked object when actual watermarked object is presented to the watermark detector module. Therefore, the amount of FNR should be negligible for real watermark system.

1.4 Zero Watermarking

In order to provide a perfect tradeoff between imperceptibility and robustness, zero watermark technique has been proposed which is constructed watermark without embedding watermark data into the host object. For this purpose, zero watermarking utilizes some features of the host object to construct watermark data which is kept by Intellectual Property Rights (IPR) organization as a third-party agency [5].

1.5 Applications of Digital Watermarking

Several applications are known for digital watermarking in which the most common applications include content identification and management, content protection for multimedia content, forensics and piracy deterrence, forensics and piracy deterrence, content filtering, communication of ownership, document and multimedia security, authentication of content and objects, broadcast monitoring, locating content online, rich media enhancement for mobile phones, audience measurement, improved auditing, and second screen [6, 7]:

(a) Content Identification and Management: By embedding a unique digital identity into a digital content without interfering user's enjoyment and perceive, there is possibility to manage the digital content through wide range of digital devices such as computers, networks, and video players. For example, the watermark can be detected in background program to trigger some predefined actions such as linking to other rich databases or Web sites with same experiences.

(b) Content Protection for Multimedia Content: By embedding a permission watermark tag into a digital content, there is possibility to permit users to copy and watch a digital media. Since watermarking is added as a security layer on top of the digital content, managing unauthorized access and protecting digital content from distribution are more achievable. For example, there is always possibility to mute audio file, show explanatory message, stop copying, and stop playback an unauthorized digital contents.

(c) Forensics and Piracy Deterrence: By embedding a situational metadata including recipient IP address, received format, transmission time, and distinct forensic watermark into a digital content at only one stage or each stage of content distribution, there is possibility to retrieve the forensic watermark as evidence. Furthermore, it is possible to discover the source of the leak to trigger contractual provisions and perform some legal actions for criminal cases.

(d) Content Filtering (Includes Blocking and Triggering of Actions): Upon content identification application, there is possibility to filter digital content. For example, a specific action (press a button for more information) can be triggered when a sense of movie is watching which allows personal interactivity. In addition, other proportion, advertisement, and commercial information can be appeared at specific times within the move which is being watched. Similarly, the content can be blocked for copyright and children safety, and to prevent specific sense to be watched by children and warn the adults which are extremely useful for the Internet content supervision.

(e) Communication of Ownership and Copyrights: By embedding watermark owner chooses such as contact details, ownership information, and usage rights into a digital content, there is possibility to track digital assets through Internet. In contrast to metadata that can be lost under file transformations and manipulations, the watermark is always tied to digital content as content unique identity in order to prove ownership and copyright.

(f) Document and Multimedia Security: By embedding a unique digital ID into a confidential document by brand owners and factories, there is possibility to be confidential about source of leak. Also, companies can put watermark detector inside scanners and printers to prevent confidential documents to be distributed and copied. Moreover, this watermark can trigger a notification e-mail for real owner.

(g) Authentication of content and objects (includes government ids): Nowadays, by embedding an authorization ID into a digital object by governments, there is possibility to combat fraud, theft, counterfeiting, and tampering of the identity. For instance, many of the driving licenses in the USA are applying digital watermarking technology to authenticate IDs quickly by detecting the presence and absence of the watermark and then compute the cross-correlation between the extracted watermark and other information on the card which improves the security by minimizing tampering and spoofing the driving licenses' identity. In other countries, digital watermarking technology is using for National ID cards which are widely used for various purposes such as commercial transactions, banking, driving, traveling, and credit card. These wide applications are due to ability of the watermark to embed imperceptible information into all digital object forms simply and effectively. Moreover, digital watermarking technology can provide reliable inspection on ID authentication in such a way that citizen privacy data are also protected.

(h) Broadcast monitoring: By embedding disseminated data into a digital content during production time and when it is broadcasting, there is possibility to prove the owner of the content. Furthermore, putting hardware or software watermark detection can reveal some information including who, how long, when, and where the content is broadcast. Moreover, for broadcast monitoring, it is possible to embed a metadata, which includes author, type of content, and title, or embed a reference entity which is linked to a database with more complete metadata associated with the content. Therefore, while the watermark is extracted from the content for rapid analyzing, the extracted broadcasting details are reported and confirmed from radio and TV stations.

(i) Locating content online: By embedding a watermark ID into a digital content by author, there is possibility to search for uniquely watermarked content through the Internet. While the Web pages are constantly crawl for the unique watermark, any result can be reported for notifying the owner of the content for necessary action.

(j) Rich media enhancement for mobile phones: By embedding a digital ID into the all forms of media content such as brochures, posters, newspapers, packaging, magazines, and more, there is possibility for smart phone to detect it and match it to a URL in a backend database that is then returned to the consumers' phone. As a consequent, there are new experiences on proprietary content for paid subscribers, promotions, discount, video, and games that should be considered. In contrast to QR codes and 2D barcodes that take up precious space on printed materials, digital watermark is brand friendly and imperceptible for human, which provide an endless opportunity for marketers and brands to engage consumers with rich and new experience through their smart phones.

(k) Audience measurement: By embedding a digital data each program's audio/video tracks during airing time, there is possibility to measure audience by installing audiometers in panelist's home, and then collect and transfer information to a central database for daily reporting and processing. Thus, digital watermarking technology enables the broadcasters to identify, analyze, and retrieve the contents instantly by specialized hardware and software.

(l) Improved auditing: By embedding an identifier for every licensed asset by distributors, there is possibility to convey any licensee compositions. Therefore, any usage of the owner's asserts including all or part of the original asset can be audited automatically and quickly by digital watermarking technology.

(m) Second screen: By embedding watermark into the audio aspects of the video content, there is possibility for Second Screen App to identify the program which is displaying and watching. The extracted information from the recorded audio by the microphone of the smart devices is rich and meaningful for second screen technology in terms of celebrity news, augmentation or enhancement, previous episode information, advertising, and merchandising to help set context and a host of other information.

Apart from the above applications, digital watermarking technology can have complemental contributions on other technology including network communication where network stream, e.g., VoIP, can be manipulated by impostors [8–10], military communication when every command should be guaranteed for originality [11–14], social security monitoring for telephonic recording, enhancing the security of online biometrics verification or identification systems [13, 14], and identifying the airplane through Very High-Frequency (VHF) radio channel for Air Traffic Control (ATC) application [15–17].

1.6 Book Structure

The remainder of this book is organized based on four parts. The first part of this book is dedicated to one-dimensional signal watermarking. This part is divided into two chapters which discuss audio and speech watermarking, respectively.

Chapter 2 of this book provides an overview on audio watermarking techniques. Firstly, basic and fundamental of audio watermarking is discussed.

Secondly, a brief exploration of audio watermarking techniques, problems, and solutions is explained. Furthermore, this chapter discusses the details in audio watermarking attacks and different metrics for evaluation. Lastly, a basic theory, related, and pervious works in audio watermarking techniques are explained.

Chapter 3 presents speech watermarking approaches with more details. The first subsection discusses the overall introduction of the speech watermarking. The second subsection describes the techniques and how it can be applied for speech watermarking properly. The third subsection presents state of the arts in speech watermarking technique. The fourth subsection discusses several advantages and disadvantages of speech watermarking techniques.

The second part of this book is dedicated to multi-dimensional signal watermarking. This part is divided into three chapters which discuss image, video, and 3D watermarking, respectively.

Chapter 4 of this book provides an overview on image watermarking techniques. Firstly, a background of image watermarking is discussed. Secondly, a brief exploration of image watermarking techniques, attacks, and challenges is explained. Furthermore, this chapter discusses the details of HVS for image watermarking. Finally, image watermarking techniques are extended for medical image watermarking purpose.

Chapter 5 of this book provides an overview on video watermarking techniques. Firstly, basic and fundamental of video watermarking is discussed. Secondly, a brief exploration of video watermarking techniques, problems, and solutions is explained. Furthermore, this chapter discusses the details of various video-enforced strategies. Lastly, a critical review in video watermarking techniques is discussed.

Chapter 6 of this book provides an overview on 3D watermarking techniques. Firstly, modeling and representation of a 3D object is explained. Secondly, a brief exploration of 3D watermarking techniques and attacks is discussed. Lastly, this chapter evaluates different 3D watermarking techniques.

The third part of this book is dedicated to document watermarking. This part is divided into four chapters which discuss natural language, software, database, and text watermarking, respectively.

Chapter 7 of this book provides an overview on NL watermarking techniques. Firstly, basic and fundamental of NLP is discussed. Secondly, a brief exploration of NL watermarking techniques, limitations, and issues is explained. At the end, difference between NL watermarking and text watermark technologies is explained.

Chapter 8 of this book provides an overview on text watermarking techniques. Firstly, basic in text watermarking is discussed. Secondly, text watermarking techniques, attacks, and criteria are explained. Lastly, a discussion in text watermarking techniques is discussed.

Chapter 9 of this book provides an overview on software watermarking techniques. Firstly, basic and fundamental of software watermarking is discussed. Secondly, formal representation of software watermarking is provided. Furthermore, this chapter discusses the details in software watermarking criteria, technique, and different attacks. Lastly, discussion and comparison on different software watermarking techniques are explained.

Chapter 10 of this book provides an overview on relational database, XML, and ontology watermarking techniques. Firstly, basic and fundamental of database watermarking is discussed. Secondly, different techniques for database watermarking, issues, attacks, and types of watermark data are explained. Furthermore, this chapter also extends the relational database watermarking for XML watermarking. Finally, a brief extension of relation database watermarking for ontology watermarking is discussed.

The fourth part of this book is dedicated to advanced topics in watermarking technology. This part is divided into three chapters which discuss network stream watermarking, hardware IP watermarking, and security enhancement in digital watermarking in watermarking technology, respectively.

Chapter 11 of this book provides an overview on network stream watermarking techniques. Firstly, network traffic modeling and network watermark properties are described. Secondly, various network stream watermarking techniques are explained. Furthermore, this chapter discusses different types of adversary and different watermarking approaches, and various attacks in network stream are described.

Chapter 12 of this book provides an overview on hardware IP design watermarking techniques. Firstly, background in IP design and System-On-Chip are explained. Secondly, various deliverable IP design levels are explained. Furthermore, this chapter discusses different techniques and attacks, and criterion in IP block is described. Lastly, a comparison and discussion on IP watermarking techniques are provided.

Chapter 13 of this book provides an overview on security enhancement in watermarking techniques. In this chapter, combination of watermarking with cryptography is discussed in order to improve the security. Moreover, different applications of the digital watermarking technology are discussed for online biometric recognition systems. The recent approach based on quantum watermarking is also explained. Finally, the differences between security and robustness for digital watermarking are described.

References

1. Nematollahi, M.A., and S. Al-Haddad. 2013. An overview of digital speech watermarking. *International Journal of Speech Technology* 16(4): 471–488.
2. Mat Kiah, M., et al. 2011. A review of audio based steganography and digital watermarking. *International Journal of Physical Sciences* 6(16): 3837–3850.
3. Rec, I., P. 800. 1996. *Methods for subjective determination of transmission quality.* International Telecommunication Union, Geneva.
4. McLoughlin, I. 2009. *Applied speech and audio processing: with Matlab examples.* Cambridge: Cambridge University Press.
5. Zhou, Y., and W. Jin. 2011. A novel image zero-watermarking scheme based on DWT-SVD. In *2011 International Conference on Multimedia Technology (ICMT)*. IEEE.

6. Gopalakrishnan, K., N. Memon, and P.L. Vora. 2001. Protocols for watermark verification. *IEEE Multimedia* 8(4): 66–70.
7. Digital Watermarking Alliance.
8. William, S. 2006. *Cryptography and network security,* 4 edn. Pearson Education India.
9. Huang, H.-C., and W.-C. Fang. 2010. Metadata-based image watermarking for copyright protection. *Simulation Modelling Practice and Theory* 18(4): 436–445.
10. Huang, H.-C., et al. 2011. Tabu search based multi-watermarks embedding algorithm with multiple description coding. *Information Sciences* 181(16): 3379–3396.
11. Faundez-Zanuy, M., J.J. Lucena-Molina, and M. Hagmüller. 2010. Speech watermarking: an approach for the forensic analysis of digital telephonic recordings. *Journal of Forensic Sciences* 55(4): 1080–1087.
12. Faundez-Zanuy, M. 2010. Digital watermarking: new speech and image applications. *Advances in nonlinear speech processing*, 84–89.
13. Faundez-Zanuy, M., M. Hagmüller, and G. Kubin. 2006. Speaker verification security improvement by means of speech watermarking. *Speech Communication* 48(12): 1608–1619.
14. Faundez-Zanuy, M., M. Hagmüller, and G. Kubin. 2007. Speaker identification security improvement by means of speech watermarking. *Pattern Recognition* 40(11): 3027–3034.
15. Hagmüller, M., et al. 2004. Speech watermarking for air traffic control. *Watermark* 8(9): 10.
16. Hofbauer, K., G. Kubin, and W.B. Kleijn. 2009. Speech watermarking for analog flat-fading bandpass channels. *IEEE Transactions on Audio, Speech, and Language Processing,* 17(8): 1624–1637.
17. Hofbauer, K., H. Hering, and G. Kubin. 2005. Speech watermarking for the VHF radio channel. In *Proceedings of the 4th Eurocontrol innovative research workshop*.

Part I
Signal Watermarking

Chapter 2
Audio Watermarking

2.1 Introduction

Audio watermarking is a well-known technique of hiding data through audio signals. It is also known as audio steganography and has received a wide consideration in the last few years. So far, several techniques for audio watermarking have been discussed in literature by considering different applications and development positions. Perceptual properties of human auditory system (HAS) help to hide multiple sequences of audio through a transferred signal. However, all watermarking techniques face to a problem: a high robustness does not come with a high watermark data rate when the perceptual transparency parameter is considered as fixed. Furthermore, selection of a suitable domain, cover, and considering the problems associated with data-hidden techniques must be considered for designing the path to achieve a data-hidden purpose.

The remainder of this chapter is organized as follows: Transmission channel for audio watermarking is discussed. Different audio watermarking attacks are explained. Various audio watermarking techniques are compared.

2.2 Transmission Channel

A signal travels from different transmission environment during its journey from transmitter to receiver. As schematically illustrated in Fig. 2.1 [1], there are four classes of transmission environments. They are digital, resampling, analog, and on the air environments.

A signal passes through a digital end-to-end environment that is the way from which a digital file is copied from a machine to another one, with no further modifications and the same sampling at the encoder and decoder. For these reasons,

© Springer Science+Business Media Singapore 2017
M.A. Nematollahi et al., *Digital Watermarking*, Springer Topics
in Signal Processing 11, DOI 10.1007/978-981-10-2095-7_2

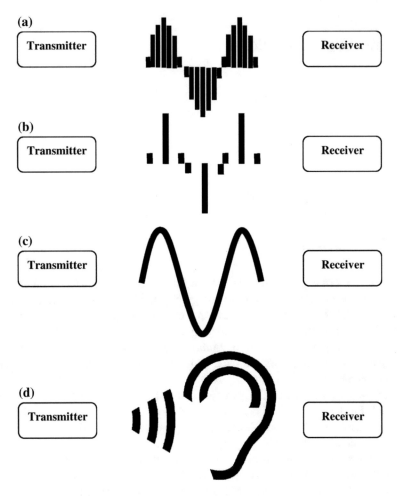

Fig. 2.1 Various transmission channels including: **a** digital, **b** resample **c** analog, and **d** over the air

the least data hidden can be applied in this class. Resampling is the second class of environment for a signal. The sampling rate for a signal during resampling is not necessary the same as its first sampling rate and temporal characteristics of the signal are subject to some modifications. Nevertheless, the signal remains in digital form throughout its way and almost the magnitude and phase of the signal remains intact. When a signal is played in analog environment, its phase is generally preserved. But some of its features do not hold their initial values, e.g., absolute signal magnitude, sample quantization, and temporal sampling rate. There is a final class on environment that is met when a signal is played on the air and is resampled with a microphone. The fact is that the signal can be modified in a nonlinear manner in terms of phase, amplitude, and frequency components (e.g., echoes). Due to different impacts of transmission environments on the characteristics of a signal and data-hiding method, it is necessary to consider all the possible environments that a signal may pass.

2.3 Audio Watermarking Techniques

Generally, many audio watermarking techniques have been developed. The well-known methods of audio watermarking based on the limitations of perceptual properties of HAS are including simple least significant bits (LSB) scheme or low-bit encoding, phase coding, spread spectrum, patchwork coding, echo coding, and noise gate technique.

A pathway for watermarking especially for the famous patchwork algorithm was proposed in [2]. His method improves the performance of the original patchwork algorithm. Another method called as modified patchwork algorithm (MPA) [3] enhanced the power of Arnold's algorithm and improved its performance in terms of robustness and inaudibility. A mathematical formulation has also been presented that aids to advance the robustness.

Spread-spectrum technology has been utilized in audio watermarking in [1] which was originally introduced in [4]. Another method based on the spread-spectrum technology in [5] is a multiple echo technique that replaces a large echo into the host audio signal with multiple echoes with different offsets. Next method is the positive and negative echo-hiding scheme [6]. Each echo contains positive and negative echoes at adjacent locations. In the low-frequency band, the response of positive and negative echoes forms a smooth shape that is resulted by similar inversed shape of a negative echo with that of a positive echo. When positive and negative echoes are employed, the quality of the host audio is not obviously depreciated by embedding multiple echoes.

Backward and forward kernels are employed in an echo-hiding scheme presented by Kim and Choi [7]. They theatrically provided some results showing that the robustness of echo-hiding scheme improves by using backward and forward kernels. They showed that when the embedded echoes are symmetric, for an echo position associated with a cepstrum coefficient, the amplitude in backward and forward kernels is higher than when using the backward kernel.

Time-spread echo kernel is then proposed by Ko et al. [8]. A pseudo-noise sequence acts as a secret key that spreads out an echo as numerous little echoes in a time region. This secret key is then applied for extraction of the embedded data of the watermarked signal. The usage of the pseudo-noise sequence is essential, because the extraction process of a watermarked audio signal becomes very though with no secret key.

In this part, the available audio watermarking techniques are divided into the three major categories. Three categories for audio watermarking are summarized in Fig. 2.2 which are based on prominent domains for embedding data in an audio signal: temporal, frequency and coded domains. In the reminder of this chapter, each method is summarized and their advantages and disadvantages are discussed.

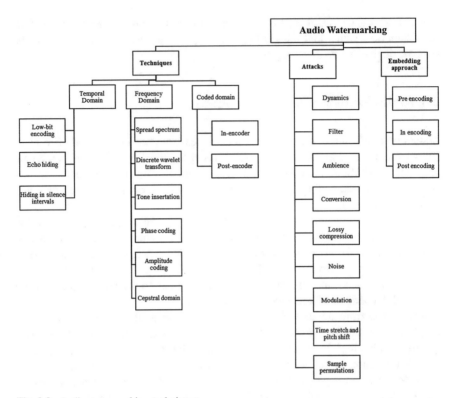

Fig. 2.2   Audio watermarking techniques

2.3.1 *Temporal Domain*

Audio watermarking techniques based on temporal domain are summarized in this section. Famous techniques for temporal domains are including low-bit encoding, echo hiding, and hiding in silence interval. In the following, each technique is fully discussed in detail.

2.3.1.1 Low-Bit Encoding

The most applied method for data hiding is called as low-bit encoding or lease significant bit (LSB) [9]. Basically, the least significant bit of the cover audio is utilized for embedding each bit from the message. For example, 8 kbps data are hidden in a signal with 8 kHz sampled audio which has 8 bits per sample. This method is relatively simple and has a high capacity for hiding data. The robustness of this method is increased when it is combined with other watermarking methods. Nevertheless, the low-bit encoding method is sensitive to noises, which reduces the security and robustness. The position of hidden data in the watermarked signal is known which makes this method vulnerable to attacks and an attacker via elimination of entire LSB plane can easily discover a message or destroy the watermark.

Basic LSB has been performed for transmission of an audio signal on a wireless network in [10]. The results verified that the method reduces the robustness and security at high rate of embedding data, but it does not harm the imperceptibility of final signal.

A method for embedding four bits per sample was presented in [11] that enhanced the hiding capacity. This method reduces the impact of error on the watermarked audio signal by defusing the embedding error on the next four samples. The depth of embedding layer of data increased from 4 layers to 6 and 8 LSB layers with no significant effect on the imperceptibility of the audio signal [12]. The results showed that the methods with higher embedding layer enhanced the robustness of previous method when noise addition and distortion occurs.

In [13], bits of the message are replaced with the bits at the sixth position of each 16-bit sample of the original audio signal. An approach for reducing the embedding error is replacing the message bits in such a way that the resulted bit sequence becomes closer to the original one. For this purpose, other bits are permitted to be flipped for increasing the closeness of bit sequence to the original one. As an instance, if four bits "0100" (value 4) are used for embedding data and bit "1" must be embedded in the bit sequence, it is suggested to select the bit sequence of "0011" (that is value 3) instead of having "1100" (that is value 12). The reason is that value 3 is closer to value 4 and the result is a lower embedding error rate.

The other approach in [12] suggests an eight layer for LSB embedding. In order to enhance imperceptibility of watermarked signal, the approach avoids hiding data in silent periods of the original signal. Due to assigning 8 bits for LSB embedding, the hiding capacity of the result becomes lower than the previous methods. However, it improves the robustness. The major disadvantage of embedding data in 6th or 8th position of LSB is the difficulty to reveal the original audio signal especially when the bits are shifted or flipped to enhance the embedding error rate.

2.3.1.2 Echo Hiding

An audio effect is known as echo which repeats some parts of the sound by creating delay inside the audio signal. In order to hide an echo, echo-hiding method generates a short echo by using a resonance and adds the echo to the original audio signal. The addition of the short echo is not recognizable by HAS; therefore, this method is not sensitive to noise addition. Other perceptual and statistical properties of original signal are kept in resulted signal.

Three parameters of the echo signal are the candidates for hiding the data. They include the initial amplitude, the delay (or offset), and the decay rate. The data can be successfully hidden in the audio signal if their values are managed to keep the imperceptibility of audio signal [14]. For this reason, the values of amplitude and decay rates should be set below the audible threshold of HAS. As an example, when the time difference between the original signal and the echo stays below 1 ms, there is no annoying effect on the audibility of the signal.

Due to the induced size of echo signal, low embedding rate, and security, there are few systems and applications that practically developed this method. To the best of our knowledge, there is no real system that uses echo hiding in audio watermarking which cannot provide sufficient data for evaluation. An echo-hiding-time spread technique has been introduced to resolve the low robustness of echo-hiding technique in facing with common linear signals [15]. This method spreads the watermark bits all over the original signal and the destination recovers them by using the correlation amount. As a result of being a cepstral content-based method, the cepstral portion of error is detached and the detection rate at the decoder gets higher.

2.3.1.3 Hiding in Silence Intervals

Another candidate for embedding data is silence intervals in speech signal. A simple approach for hiding in silence intervals is proposed [16]. Consider n as the number of required bits for denoting a value from the message to hide. The silence intervals in audio signal should be detected and measured in terms of the number of samples in a silence interval. These values are decremented by x, $0 < x < 2n$ bits, where $x = \text{mod}(\text{new\_interval\_length}, 2n)$. As an instance, consider that the value 6 is hided in a silence interval with length 109. Taken 7 samples out from the interval, 102 samples are remained in the new interval. The value x is computed as $x = \text{mod}(102, 8) = 6$. The short length of silence intervals that commonly seen in continuous parts of normal audios is omitted from the portions for hiding data. The perceptual transparency of this method is acceptable, but compression of signal misleads the data extraction process. As a solution for this problem, an approach is presented in [17] which separates the silence intervals from audio intervals so that they are not interpreted as one another. Thus, it reduces the samples in silence intervals and slightly augments the samples of the audio interval. The first and last interval added to the audio during MP3 coding is simply ignored in data hiding and retrieval.

As a general conclusion, conventional LSB approach is simpler than other methods; however, its capacity for hiding data is low. Moreover, it is resilient to noise additions and shows higher robustness in comparison with its variants [12, 13]. The main difficulty is a few number of applications that use time domain techniques.

2.3.2 Frequency Domain

Main idea behind using the frequency domain (or transform domain) for hidden data is the limitation of HAS when frequency of an audio signal fluctuates very rigid. The "masking effect" phenomenon enables the HAS to mask weaker frequency near stronger resonant frequencies [18]. It provides a time duration that can be utilized for embedding data. The data hidden in this space is not perceptible by HAS. Watermark methods in frequency domain directly manipulate the masking effect of HAS by explicit modification of masked regions or indirectly by slight change of the samples of the audio signals.

2.3.2.1 Spread Spectrum

By spreading data in the frequency domain, spread spectrum (SS) technique ensures an appropriate recovery of the watermarked data when communicated over a noise-prone channel. SS utilizes redundancy of data for degrading the error rate of data hiding. An M-sequence of code handles the data and is embedded in the cover audio. This sequence is known to sender and receiver and if some parts of these values are modified by noise, recovery of data is feasible by using other copies [19]. The SS technique was developed in MP3 and WAV signals for the purpose of hiding confidential information in the form of conventional direct-sequence spread spectrum (DSSS) technique [20].

A frequency mask was suggested for embedding the data in a watermarked audio signal [21]. When a phase-shifting approach is combined to SS, the result is a watermarked signal with a higher level of noise resistance and robustness. As discussed in [21], the detection of hidden data is simple in the new method, but the rate of hiding data is low. As a solution, sub-band domain is chosen to provide better robustness and improving the decoder's synchronization uncertainty which require to select proper coefficients in sub-band domain [22].

2.3.2.2 Discrete Wavelet Transform

Discrete wavelet transform (DWT) is multi-scale and multi-resolution technique to decompose signal to different time-frequency components. A watermarking method is proposed by DWT which hides data in LSB of the wavelet coefficients [23]. The imperceptibility of hidden data is low in DWT. Whenever the integer wavelet coefficients are available, a hearing threshold is useful to improve the audio inaudibility as presented in [24]. If a DWT watermarking technique evades embedding data in silent parts, hidden data does not annoy the audience [25]. DWT provides a high rate of data hiding; nevertheless, the procedure for data extraction at the receiver is not always accurate.

2.3.2.3 Tone Insertion

HAS does not detect audio signals when lower power tones are located near very high tones. Tone insertion benefits this HAS feature for data hiding. The method to embed inaudible tones in cover signal was introduced in [26]. Given that one bit is planned to be hided in an audio frame, two frequencies of f_0 and f_1 are selected and a pair of tones is created in this area. Each frequency has a masked frequency, e.g., pf_0 for f_0 and pf_1 for f_1. Considering there are n frames and the power of each frame is denoted by pi where $i = 1,\ldots, n$. The value of each masked frequency is set to a predefined value that is the ratio of the general power of each audio frame pi. A correct data extraction from watermarked data is obtained when tones are inserted at known frequencies and at low power level.

Procedure of detection of the hidden data from the inserted tones is performed by computing the power of each frame, pi, including the power of pf_0 for f_0 and pf_1 for f_1. If the ratio $pi\, pf\, 0 > pi\, pf\, 1$, then the hidden bit is assumed as "0"; otherwise, it is considered as "1." Thus, the hidden data is extracted. As perceived, the data-hiding capacity of tone insertion method is low. Some attacks can be tolerated by tone insertion method, e.g., low-pass filtering and bit truncation; nonetheless, the attackers can simply detect the tones and extract the hidden data. Similar to LSB, this problem can be resolved by varying four or more pairs of frequencies in a keyed order.

2.3.2.4 Phase Coding

Another limitation of HAS is its inability to detect the relative phase of different spectral components. It is the basis of interchanging hidden data with some particular components of the original audio signal. This method is called as phase coding and works well on the condition that changes in phase components are retained small [27]. Phase coding tolerates noises better than all other above-mentioned methods [1, 28].

An independent multi-band phase modulation is utilized for phase coding [27]. In phase modulation method, phase alteration of the original audio signal is controlled to obtain imperceptibility of phase modifications. Phase components are determined by quantization index modulation (QIM). Then, the nearest "o" and "x" points are replaced with phase values of frequency bin to hide "0" and "1," respectively. Therefore, phase coding achieves a higher robustness when perceptual audio compression is applied [1].

QIM was widely been used that improves the capacity of data hiding of phase coding by replacing the strongest harmonic with step size of $\pi/2n$ [29]. Phase coding has zero value of bit error rate (BER) when MP3 encoder is applied that demonstrates the high robustness of this method.

As HAS is not sensitive to phase changes, an attacker simply can replace his/her data with the real hidden data. S/he can apply frequency modulation in an inaudible way and modify the phase quantization scheme.

2.3.2.5 Amplitude Coding

The sensitivity of HAS is high for frequency and amplitude components. Therefore, it is possible to embed hidden data in the magnitude audio spectrum.

The capacity of hiding data is high by using this method as presented in [28] and the tolerance of the method regarding noise distortion and its security in facing with different attacks is high. Hiding different types of data is feasible by using this method. Encrypted data, compressed data, and groups of data (LPC, MP3, AMR, CELP, parameters of speech recognition, etc.) can be hided by using amplitude coding.

Initially, some spectrum areas for secure embedding data are found in the wideband magnitude audio spectrum. For this purpose, an area below 13 dB of the original signal spectrum is taken into account and a frequency mask is defined in this area. In regard to the magnitude spectrum, a distortion level that is resilient to noise distortion is considered. Then, candidate locations and the capacity for hiding data can be determined.

For 7 to 8 kHz frequencies, the effect on the wideband speech is minimum [30]. Therefore, this area is a good space for hiding data with not compromising the inaudibility of watermarked signal. For this purpose, the entire range between 7 and 8 kHz can be filled with hidden data.

2.3.2.6 Cepstral Domain

Cepstrum coefficients provide spaces for watermarking. This method is resilient to well-known attacks in signal processing and is also known as log-spectral domain. It locates the hidden data in the portions of frequencies that are inaudible by HAS and obtains a high capacity of hiding data, between 20 and 40 bps [31]. Initially, the domain of original audio signal is modified to cepstral domain. Statistical mean function helps to choose some cepstrum coefficients that are later altered by hidden data. As the masked regions of the majority of cover audio frames are utilized for data hiding, the imperceptibility of watermarking is relatively high in cepstral domain.

The robustness of this method was improved by considering high energetic frames and replacing cepstrum of two selected frequencies F_u and f_2 by bit "1" or "0" [32]. The security and robustness of this method was later improved by considering different arbitrary frequency components at each frame [33]. Distinct types of all-pass digital filters (APF) choose sub-bands that are suitable for embedding hidden data. A hiding method based on APF improves the robustness of watermarked audio signal facing with addition of noise, random chopping, e-quantization, and resampling [34]. Given n as an even positive integer, the robustness can be further improved by applying a set of n-order APFs as is in [35]. Pole locations of an APF are calculated from the power spectrum by several approaches. Finally, the data is hidden in some chosen APF parameters.

According to calculations, all the above-mentioned techniques have higher resilience against noise additions in frequency domain (or transform domain) [28]. Almost all data-hiding methods in transform domain benefits the perceptual models of HAS, especially frequency masking effect, to improve the data-hiding capacity as long as signal distortion can be tolerated. Most of the watermarking methods in transform domain is tolerating simple noise distortions including amplification, filtration, or resampling. However, the probability of them to tolerate noisy transmission environment or data compression in ACELP and G.729 is low.

2.3.3 Coded Domain

In real-time communications, coded domain is favorable. Despite the benefits of transform domain in comparison with time domain, it does not act well when real-time applications and voice encoders under particular encoding rates, e.g., AMR, ACELP, and SILK, are employed. An encoder codes the audio signal while it is transferring through communication channels and at the end, a decoder is responsible for decoding the coded data. As the encoder and decoder have their own rates, a decoded signal might slightly differ the original signal. Therefore, the procedure for data extraction and retrieval is complicated in coded domain. Furthermore, the correctness of the extracted data is a challenge itself.

2.3.3.1 In-Encoder Techniques

A coded technique called as in-encoder technique was introduced that can successfully tolerate noise distortion, audio codec, compression, and reverberations [36]. Different types of audio signals including music and speech were evaluated for embedding watermarked data when sub-band amplitude modulations have been used.

A pitch-tracking algorithm based on autocorrelation performed voiced/unvoiced segmentation in [37] based on the LPC vocoder. A data sequence was embedded in the unvoiced segments by alteration of the linear prediction residual. This method does not affect the audibility of the watermarked signal if the residual's power is matched. Capacity of a reliable data hiding is up to 2 kbps. Hidden data is replaced with the unmodified coefficients of the LPC filter, and for decoding the embedded data, a linear prediction analysis on the transmitted audio signal is perfumed.

A coded technique that hides the data in the audio codecs and in the LSB of the Fourier transform was proposed in [18]. This technique embeds data in the LSB of the Fourier transform of the prediction residual of the host audio signal. This technique does not guarantee inaudibility of watermarked data and its imperceptibility is considered as low. It automatically shapes the spectrum of LSB noise when an LPC filter is employed; thus, the watermarked data has a less impact on audibility of the audio signal.

2.3.3.2 Post-encoder Techniques

The watermark can be embedded in the coded domain by the post-encoder (or in-stream) techniques. A post-encoder technique was developed on an AMR encoder at a rate of 12.2 Kbit/s and in the bitstream of an ACELP codec [38]. It works together with the analysis-by-synthesis codebook search and the results showed that it hides 2 Kbit/s of data in the bitstream and obtains a noise ratio of 20.3 dB. A lossless post-encoder technique was developed that works on G.711-PCMU telephony encoder [39]. Data is presented in the form of folded binary code. The value of each sample varies between -127 and $+127$ (consists of values -0 and $+0$). For every 8-bit sample with absolute amplitude of zero, one bit is hided. Thus, the capacity of hidden data varies between 24 and 400 bps. As a solution for improving capacity of hidden data for G.711-PCMU, a semi-lossless approach was proposed in [40]. A predefined level, denoted as "i," amplifies the sample's amplitudes. Hereafter, the samples with absolute amplitude between 0 and i are applied for embedding data. For increasing the capacity of watermarked data, in [41] the inactive frames in low-bit-rate audio stream (i.e., 6.3 kbps) were used for encoding a G.723.1 source codec.

In general, coded domain techniques are well suited for real-time applications. Watermarking techniques especially in-encoder approaches benefits from a high robustness and security. While capacity of hidden data is higher than the codec data in some techniques; due to high sensitivity of bitstream to modifications, it is held small to limit the perceptibility. Although ACELP, AMR, or LPC audio codecs and noise additions are tolerable by coded domain techniques, the integrity of hidden data cannot be promised where transcoding (i.e., a voice encoder/decoder) is available in the networks. A voice enhancement method that is applied for reducing the noise or echo can modify the hidden data, as well. However, the procedure of data extraction in tandem-free operation guarantees that hidden data remains intact during encoding the data encoding.

2.4 Embedding Approach

In covert communication, data is transferred through multiple encoders/decoders. An encoder reduces the size of transmitted data by removing the redundant or unused data. Thus, each coder influences the integrity of data, while the robustness of covert communications requires a high integrity of watermarked data. Although, there are some ways to ensure data integrity in encoder/decoder, it imposes negative impacts on hiding capacity of data. There are three levels for embedding a data-in-audio watermark system [38]. Figure 2.3 summarizes the aforementioned methods for audio steganography according to the occurrence rate. The evaluation of security requires a third-party effort cost to retrieve the hidden data. Each level has some benefits and weaknesses that are discussed as follows.

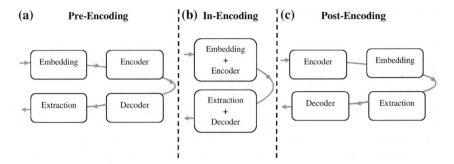

Fig. 2.3 Different approaches for embedding the watermark

2.4.1 *Embedding Before Encoding (Pre-encoding)*

Prior to encoding process, the data is embedded in time and frequency domain. This level is known as pre-encoder embedding. The integrity of data, during transmission over network, is not guaranteed in this level because high degree of data compression in encoders (e.g., in ACELP or G.729) and addition of noise (in any form, e.g., WGN) can compromise the integrity of data. On the other hand, there are some methods that allow a low degree of modifications on the audio signal including resizing, resampling, filtering. Therefore, they are resilient to low degree of noise addition or data compression. Only noise-free environments provide a space for high rate of data hidden.

2.4.2 *Embedding During Encoding (in-Encoder)*

This data embedding level provides a robust data hiding. For this purpose, a codebook of codecs is necessary. The codebook keeps the information of transmitted data once the requantization operation is performed. As a result, for every parameter of audio signal, two important values of embedded-data and codebook parameters are kept. When value of embedded data is manipulated for any reason, this method faces to a severe problem for data extraction. It can occur when the data passes through a voice encoder/decoder in a radio access network (BST, BSC, TRAU) and/or in the core network (MSC) in a GSM network. Similar modifications occur when a voice enhancement algorithm is developed in a radio access network and/or in the core network.

2.4.3 Embedding After Encoding (Post-encoder)

This level of embedding data acts on bitstreams rather than the original audio signal. Data is hidden in a bitstream once it passes the encoder and before entering the decoder. Thus, value of data and the integrity of watermarked audio signal are vulnerable to undesirable modifications. Bitstreams are naturally more sensitive to alteration than audio signals and data integrity should be kept small to avoid imperceptibility of audio signal. Nevertheless, post-encoder embedding ensures the correctness of data once it is extracted in tandem-free operations and the message is retrieved in a lossless way.

2.5 Audio Attacks

As shown in Fig. 2.2, there are many attacks that can degrade the watermark data and as a consequence decrease the robustness of the audio watermarking techniques. Some of the attacks have already discussed in the literature for still images and some have been particularly mentioned for audio watermarking. In this section, the impact of each attack according to the audibility of hidden data by HAS is measured and the most effective attacks on audio signals are highlighted. Some of the attacks mostly occur in real environments. Suppose an audio signal is prepared to be broadcast on a radio channel. Based on the audience confidence and quality parameters of the radio channel, the audio material is normalized and compressed to fit the necessary level of loudness for transmission. Then, the quality of signal is optimized by equalization; undesired parts are demised or dehisced; useful frequencies are kept and unnecessary ones are omitted by filters.

In some applications, the robustness of watermarked audio signal should be high, e.g., in commercial radio transmission or copyright protection of music. In both examples, the watermark technique should not allow the signal to be destroyed or manipulated by attackers and if an attack occurs, it should not allow the attacker to misuse or reuse the signal. A well-known attack in this situation is lossy compression in MP3 at high rate of compressions. In addition to individual attacks, some attacks act in the form of groups. The group of attacks is also taken into account for performance evaluation of watermarking techniques. Main group attacks are including dynamics, filter, ambience, conversion, loss comparison, noise, modulation, time stretch (pitch shift), and sample permutation.

2.5.1 Dynamics

This group of attacks influences the loudness profile of an audio file. Some attacks including increasing or decreasing are simple and considered as the basic attacks.

Some attacks perform nonlinear functions including compression, expansion, and limiting. Thus, they are complicated. In another category, frequency range or a part of that is modified by frequency-dependent algorithm.

2.5.1.1 Compressor

When it is desired to decrease the strength of a signal in terms of its range, a compressor can be utilized. It can increase the overall loudness of a signal by degrading the peaks below a particular value with no distortions. Given a fast and inaudible attack that changes all signals louder than -50 dB by a small amount. It has the following properties: Attack time 1 ms, release time 500 ms, output gain 0 dB, threshold -50 dB, and ratio 1:1.1.

2.5.1.2 Denoiser

In some cases, it is essential to find a way for noise removal from the signal. Denoiser acts as a gate. It passes the eligible parts of the signal and blocks the noises. A denoiser needs a value to be used for detection of a noise. A basic denoiser simply considers loudness of signal as a noise, prior that a proper value of the loudness should be set. Here, the setting is assumed as -80 and -60 dB. Indeed, for detection of complicated noises, other techniques, e.g., DE clickers, and advanced tools are required.

2.5.2 Filter

Filters modify a spectrum by passing desired values and omitting undesired parts of the signal. Various filters have been introduced in signal processing. The basic filters are the high-pass filter and the low-pass filter. As equalizers increase or decrease some particular parts of spectrum, they can be counted as filters.

High-pass filter eliminates all frequencies below a particular value, here 50 Hz.

Low-pass filter eliminates all frequencies above a particular value, here 15 kHz.

Equalizer subtracts the frequency by a particular value, here by 48 db. The used bandwidth was frequency/10.000. Three versions of this attack have been tested using a range from 31 Hz to 16 kHz: 10 frequencies with the distance of 1 octave, 20 frequencies with the distance of 1/2 octave, and 30 frequencies with the distance of 1/3 octave.

L/R-splitting is an equalizer effect that increases the supposed stereo image. It works on two channels. In one channel, the frequency shares are reduced and are increased in the other channel. 20 frequency channels divide the spectrum. For each and every second, the value of frequency on the left channel is subtracted by dB and is increased by this value on the right radio channel. Finally, the volume of both channels is normalized to cover the volume changes.

2.5.3 Ambience

Consider an audio signal broadcasting in a room. In order to simulate this condition, reverb and delay parameters assist this group. By assigning various values to each parameter, many different qualities of effects are achieved.

Delay: The original signal is duplicated, and by the addition of the copy to the original audio signal, a wide space is simulated. Here, the volume of the delayed signal is 10 % of the original one and the delay duration is 400 ms.

Reverb: For simulation of rooms or building, reverb is utilized. Although it is similar to delay, it is shorter in delay time and reflections.

2.5.4 Conversion

Depending on the application and tools, the formats of audio material are modified, e.g., to play a mono-audio material on an stereo device, data is duplicated. The sampling rate of devices has been changed from 32 to 48 kHz and now even 96 kHz or sample size changes from 16 to 24 bit and vice versa.

Resampling: Sometimes for adaptation of devices, an audio signal is resampled by a different sampling frequency from the initial one, e.g., in CD production an audio signal is downsampled from 48 to 44.1 kHz. Resampling is similar to low-pass filter when a reduction to the highest possible frequency performed, e.g., a change from 44.1 to 29.4 kHz.

Inversion: inversion changes the sign of the samples, but the changes are imperceptible. For a comprehensive evaluation of watermarking technique, this test is also taken into account.

2.5.5 Loss Compression

Some compression algorithms work based on psychoacoustic effects of audio signal. They reduce the size of the compressed data to 10 or less times of the original data size.

2.5.6 Noise

So far, several attacks have been discussed. The result of most of the attacks is a noise. As already discussed, different sources of noise are known. Hardware components are the most effective sources of noise in audio signals. There is another attack that adds noise to terminate the watermark.

Random noise: This noise is made by addition of random numbers to the samples of an audio signal. Random numbers are limited to a particular percentage of the original audio signal. It can be considered up to 0.91 % of the original sample value on the condition that it does not compromise the quality of signal.

2.5.7 Modulation

Modulation effect can be considered as attacks, but they usually do not happen in postproduction. Software for processing audio signals can include modulation attacks. They are as follows:

Chorus: Sounds from multiple resources in the form of a modulated echo is added to the original audio signal. The delay time and strength and number of voices are different. Here, 5 voices, 30 mms max. delay, 1.2 Hz delay rate, 10 % feedback, 60 ms voice spread, 5 db vibrato depth, 2 Hz vibrato rate, 100 % dry out (unchanged signal), and 5 % wet out (effect signal)are taken into account.
Flanger: when a delayed signal is added to the original signal, flanger is generated. The delay is short and the length changes constantly.
Enhancer: An audio signal becomes more brilliant or excited if the amount of high frequencies is increased. To simulate the effect of enhancer (or exciter), sound forge is applied and medium setting is used. Detailed information about the parameters is not provided by the program.

2.5.8 Time Stretch and Pitch Shift

Time stretch and pitch shifts help to fine-tuning or fitting audio into time windows by changing the length of the audio signal with no changes in the pitch or vice versa.
Pitch Shifter: A complicated algorithm for editing audio signals is pitch shifter. This algorithm changes the base frequency of the signal with no modifications in the speed. So far, multiple pitch shifter algorithms have been presented in the literature. Selection of proper algorithm depends on the expected quality of the signal. The sound forge increases the pitch by 5 cent, and this is 480th of an octave.
Time Stretch: Time stretch prolongs or shortens the duration of an audio signal with no modification on the pitch. Here, a sound that forges with a length of 98 % of the original duration is considered.

2.5.9 Sample Permutations

An uncommon way to attack watermarks hidden in audio files is sample permutation. This group consists of algorithms that permute or drop samples and are not applicable in normal environments.

Table 2.1 Comparison among various audio watermarking techniques

Watermarking domain	Technique	Description	Benefits	Drawback	Capacity
Temporal domain	Low-bit encoding	– The most applied method for data hiding – The simplest method for data hiding into data structures, e.g., data of audio in image file or data of image in audio file – Replaces LSB plane of each sampling point with hidden data – Higher embedding layer enhanced the robustness when noise addition and distortion occurs.	Simple to develop and high bit rate	Low security, sensitive to attacks, easy to intrude	16 kbps
	Echo hiding	– Embedding data in a short echo – Echo is generated by a resonance – Data hiding can be applied by three parameters of an echo signal: initial amplitude, the delay (or offset), and the decay rate. – The values of amplitude and decay rates should be set below the audible threshold of HAS – Two echoes with different offsets are utilized for embedded data: the binary datum "one" and the other to represent the binary datum "zero."	Lossy data compression is tolerated	Low security, low hidden data capacity	50 bps
	Silence intervals	– Embeds the hidden data in silence intervals of a speech audio signal	Lossy data compression is tolerated	Low hidden data capacity	64 bps
Transform domain	Magnitude spectrum	– Instead of time domain, frequency domain is utilized – It is more resilient to noises in comparison with time domain	More resilient to noise addition during communications, higher rate of data hiding	Low robustness to simple audio manipulations	20 Kbps

(continued)

Table 2.1 (continued)

Watermarking domain	Technique	Description	Benefits	Drawback	Capacity
	Tone insertion	– Embeds inaudible tones in cover signal – A correct data extraction from watermarked data is obtained when tones are insert at known frequencies and at low power level – The data-hiding capacity of tone insertion method is low. – Some attacks can be tolerated by tone insertion method, e.g., low-pass filtering and bit truncation; nonetheless, the attackers can simply detect the tones and extract the hidden data. – Security can be upgraded by varying four or more pairs of frequencies in a keyed order.	Inaudibility of hidden data	– Low transparency – Low security	250 bps
	Phase spectrum	– Hides data in a reference phase – Replaces the phase of original audio signal with a reference phase – Phase of subsequent segments is adjusted in order to preserve the relative phase between segments – Works well if changes in phase components are retained small. – Tolerates noises well	Robust against signal processing manipulation and data retrieval needs the original signal	Low rate of data hiding	333 bps
	Spread spectrum	– Spreads hidden data in frequency domain – By spreading the encoded data, encodes stream of information on as much of the frequency as possible – Utilizes redundancy of data for degrading the error rate of data hiding – If interference on some – Frequencies is existed, the signal reception is permitted	high robustness	Vulnerable to time Scale modification	20 bps

(continued)

Table 2.1 (continued)

Watermarking domain	Technique	Description	Benefits	Drawback	Capacity
	Cepstral domain	– Data is replaced with cepstral coefficients – Locates the hidden data in the portions of frequencies that are inaudible by HAS – Obtains a high capacity of hiding data – APF improves the robustness of watermarked audio signal facing with addition of noise, random chopping, e-quantization, and resampling.	Robust against signal processing operations	Perceptible signal distortions and low robustness	54 bps
	Wavelet	– Data is replaced with the coefficients of wavelet – Hides data in LSB of the wavelet coefficients – The imperceptibility of hidden data is low in DWT – Whenever the integer wavelet coefficients are available, a hearing threshold is useful to improve the audio inaudibility	High rate of data hiding	Inaccurate data extraction at the receiver	70 kbps
Coded domain	Codebook modification	– Requires a codebook – Codebook parameters are modified to hide data	High robustness	Low capacity of hidden data	2 kbps
	Bitstream hiding	– Generates a bitstream by encoding – LSB is applied on the bitstream – Data is hided in a bitstream – Bitstreams are naturally more sensitive to alteration than audio signals	High robustness	Low capacity of hidden data	1.6 kps

Zero-Cross Inserts: This attack finds value 0 in the samples and replaces them with 20 zeros. The result is a small pause in the signal. The pause length is minimum 1 s.

Copy Samples: this attack randomly selects some samples and duplicates throughout the signal. Therefore, the signal becomes longer than the original length. Here, the signal was repeated 20 times in 0.5 s.

2.6 Comparison Among Different Audio Watermarking Methods

In order to compare and classify the audio watermarking methods, some criteria must be chosen and defined. Based on the literature, major criteria for analysis and comparison of watermarking methods are considered as robustness, security, and hiding capacity (payload). Other parameters including the transmission environment and the application influence the evaluation criteria. For that, they should be considered for performance evaluation of every watermarking technique.

In an application where multiple levels of coding and decoding are planned, evaluation of a criterion like robustness is not possible without considering the environment constraints. Table 2.1 demonstrates general watermarking domains by taken into account the major techniques in each domain (the main idea is got from [28]). The details of each technique along with benefits, drawbacks, and obtained capacity of watermarking are brought in the table as well.

References

1. Bender, W., et al. 1996. Techniques for data hiding. *IBM Systems Journal*. 35(3.4): 313–336.
2. Arnold, M. 2000. Audio watermarking: features, applications, and algorithms. In *IEEE International Conference on Multimedia and Expo (II)*. Citeseer.
3. Yeo, I.-K., and H.J. Kim. 2003. Modified patchwork algorithm: A novel audio watermarking scheme. *IEEE Transactions on Speech and Audio Processing* 11(4): 381–386.
4. Cox, I.J., et al. 1997. Secure spread spectrum watermarking for multimedia. *IEEE Transactions on Image Processing* 6(12): 1673–1687.
5. Xu, C., et al. 1999. Applications of digital watermarking technology in audio signals. *Journal of the Audio Engineering Society* 47(10): 805–812.
6. Oh, H.O., et al. 2001. New echo embedding technique for robust and imperceptible audio watermarking. In *2001 IEEE international conference on acoustics, speech, and signal processing, 2001. Proceedings. (ICASSP'01)*. IEEE.
7. Kim, H.J., and Y.H. Choi. 2003. A novel echo-hiding scheme with backward and forward kernels. *IEEE Transactions on Circuits and Systems for Video Technology* 13(8): 885–889.
8. Ko, B.-S., R. Nishimura, and Y. Suzuki. 2005. Time-spread echo method for digital audio watermarking. *IEEE Transactions on Multimedia* 7(2): 212–221.
9. Chowdhury, R., et al. 2016. *A view on LSB based audio steganography.*
10. Gopalan, K. 2003. Audio steganography using bit modification. In *ICME'03. Proceedings. 2003 International Conference on Multimedia and expo, 2003*. IEEE.

11. Cvejic, N., and T. Seppanen. 2002. Increasing the capacity of LSB-based audio steganography. In *2002 IEEE workshop on multimedia signal processing*. IEEE.
12. Ahmed, M.A., et al. 2010. A novel embedding method to increase capacity and robustness of low-bit encoding audio steganography technique using noise gate software logic algorithm. *Journal of Applied Sciences* 10(1): 59–64.
13. Cvejic, N., and T. Seppanen. 2004. Reduced distortion bit-modification for LSB audio steganography. In *2004 7th international conference on signal processing, 2004. Proceedings. ICSP'04*. IEEE.
14. Gruhl, D., A. Lu, and W. Bender. 1996. Echo hiding. In *Information Hiding*. Springer.
15. Erfani, Y., and S. Siahpoush. 2009. Robust audio watermarking using improved TS echo hiding. *Digital Signal Processing* 19(5): 809–814.
16. Shirali-Shahreza, S., and M. Shirali-Shahreza. 2008. Steganography in silence intervals of speech. In *International conference on intelligent information hiding and multimedia signal processing*. IEEE.
17. Shirali-Shahreza, M.H., and S. Shirali-Shahreza. 2010. Real-time and MPEG-1 layer III compression resistant steganography in speech. *Information Security, IET* 4(1): 1–7.
18. Kang, G.S., T.M. Moran, and D.A. Heide. 2005. *Hiding information under speech*. DTIC Document.
19. Li, R., S. Xu, and H. Yang. 2016. Spread spectrum audio watermarking based on perceptual characteristic aware extraction. *IET Signal Processing*.
20. Kirovski, D., and H.S. Malvar. 2003. Spread-spectrum watermarking of audio signals. *IEEE Transactions on Signal Processing* 51(4): 1020–1033.
21. Matsuoka, H. 2006. Spread spectrum audio steganography using sub-band phase shifting. In *International conference on intelligent information hiding and multimedia signal processing, 2006. IIH-MSP'06*. IEEE.
22. Li, X., and H.H. Yu. 2000. Transparent and robust audio data hiding in subband domain. In *International conference on information technology: coding and computing, 2000. Proceedings*. IEEE.
23. Cvejic, N., and T. Seppänen. 2002. A wavelet domain LSB insertion algorithm for high capacity audio steganography. In *Proceedings of 2002 IEEE 10th digital signal processing workshop, 2002 and the 2nd signal processing education workshop*. IEEE.
24. Delforouzi, A., and M. Pooyan. 2008. Adaptive digital audio steganography based on integer wavelet transform. *Circuits, Systems and Signal Processing* 27(2): 247–259.
25. Shirali-Shahreza, S., and M. Manzuri-Shalmani. 2008. High capacity error free wavelet domain speech steganography. In *IEEE international conference on acoustics, speech and signal processing, 2008. ICASSP 2008*. IEEE.
26. Gopalan, K., and S. Wenndt. 2004. Audio steganography for covert data transmission by imperceptible tone insertion. In *Proceedings of the IASTED international conference on communication systems and applications (CSA 2004), Banff, Canada*.
27. Ngo, N.M., and M. Unoki. 2016. Method of audio watermarking based on adaptive phase modulation. IEICE transactions on information and systems 99(1): 92–101.
28. Djebbar, F., et al. 2012. Comparative study of digital audio steganography techniques. *EURASIP Journal on Audio, Speech, and Music Processing* 2012(1): 1–16.
29. Dong, X., M.F. Bocko, and Z. Ignjatovic. 2004. Data hiding via phase manipulation of audio signals. In *IEEE international conference on acoustics, speech, and signal processing, 2004. Proceedings.(ICASSP'04)*. IEEE.
30. Guerchi, D., et al. 2008. Speech secrecy: an FFT-based approach. *International Journal of Mathematics and Computer Science* 3(2): 1–19.
31. Li, X., and H.H. Yu. 2000. Transparent and robust audio data hiding in cepstrum domain. In *2000 IEEE international conference on multimedia and expo, 2000. ICME 2000*. IEEE.
32. Gopalan, K. 2005. Audio steganography by cepstrum modification. In *IEEE international conference on acoustics, speech, and signal processing, 2005. Proceedings.(ICASSP'05)*. 2005. IEEE.

33. Gopalan, K. 2009. A unified audio and image steganography by spectrum modification. In *IEEE international conference on industrial technology, 2009. ICIT 2009*. IEEE.
34. Ansari, R., H. Malik, and A. Khokhar. 2004. Data-hiding in audio using frequency-selective phase alteration. In *IEEE international conference on acoustics, speech, and signal processing, 2004. Proceedings. (ICASSP'04)*. IEEE.
35. Malik, H., R. Ansari, and A.A. Khokhar. 2007. Robust data hiding in audio using allpass filters. *IEEE Transactions on Audio, Speech, and Language Processing* 15(4): 1296–1304.
36. Nishimura, A. 2008. Data hiding for audio signals that are robust with respect to air transmission and a speech codec. In *IIHMSP'08 international conference on intelligent information hiding and multimedia signal processing, 2008*. IEEE.
37. Hofbauer, K., and G. Kubin. 2006. High-rate data embedding in unvoiced speech. In *INTERSPEECH*.
38. Geiser, B., and P. Vary. 2008. High rate data hiding in ACELP speech codecs. In *IEEE international conference on acoustics, speech and signal processing, 2008. ICASSP 2008*. IEEE.
39. Aoki, N. 2008. A technique of lossless steganography for G. 711 telephony speech. In *International Conference on Intelligent Information Hiding and Multimedia Signal Processing*. IEEE.
40. Aoki, N. 2010. A semi-lossless steganography technique for G. 711 telephony speech. In *2010 sixth international conference on intelligent information hiding and multimedia signal processing (IIH-MSP)*. IEEE.
41. Huang, Y.F., S. Tang, and J. Yuan. 2011. Steganography in inactive frames of VoIP streams encoded by source codec. *IEEE Transactions on Information Forensics and Security* 6(2): 296–306.

Chapter 3
Speech Watermarking

3.1 Introduction

Speech is the most important form of human communication which carries valuable information on who/what/how speaker speaks. Currently, applying speech signal for computer science is growing due to three major reasons [1]. First, speech is easy to be produced, captured, and transmitted as it has a lower cost compared to image. Second, speech signal can be captured from a distance (non-invasive). Third, speech carries other types of information such as emotion, age, and gender.

In recent years, communication and computer technologies are rapidly growing which allow transferring and sharing of digital speech without any limitation. Moreover, available speech editing software is able to modify just small parts of the speech signal for changing the meaning of the speech signal. In addition, speech synthesizing technology can be applied to produce the desired individual speech signal undetectable by HAS. Therefore, applying digital watermarking seems to be necessary to solve security, privacy, and protection problems. Speech watermarking as a popular and efficient is utilizing for speech signal. Recently, speech watermarking technology can contribute to other technology, e.g., VoIP [2–4], military communication to guarantee for originality [5–8], security of telephonic recording, enhancing the security of online speaker/speech recognition systems [7, 8], and ATC purpose by identifying the airplane through watermarking the VHF radio channel [9–11].

This chapter provides information about universal speech model (LPA) and some preliminary information about speech signal. Furthermore, this chapter reviews traditional approaches and related works on speech watermarking techniques to reveal the advantages and disadvantages of each technique.

© Springer Science+Business Media Singapore 2017
M.A. Nematollahi et al., *Digital Watermarking*, Springer Topics
in Signal Processing 11, DOI 10.1007/978-981-10-2095-7_3

3.2 Speech Versus Audio

Unlike the audio (music) signal which has non-stationary and non-deterministic natures, each portion of the speech signal (between 20 and 30 ms) can be modeled by linear predictive analysis (LPA) due to quasi-stationary nature [12]. In addition, speech and audio signals have different structures in terms of syntactic/semantic structure, temporal structure, and spectral structure. Furthermore, the difference between audio and speech signals lies in consonants, zero-crossing rate (ZCR), energy sequences, tonal duration, excitation patterns, harmonic pattern, dominant frequency, fundamental frequency, power distribution, alternative sequence, tonality, bandwidth, perception, and production [13]. Other difference between speech and audio signals is related to energy concentration which is less than 4 kHz and limited to 8 kHz for speech signal and is extended to 20 kHz for audio signal. It must be mentioned that these differences are used in speech/audio discriminator algorithms.

Generally, intangibility is more important than quality in speech. However, in music, quality is more important. There is a higher possibility to tolerate noise in the speech signal until intangibility happens. As a result, embedding more robust watermarks by using more watermark intensity or embedding more watermark bits is justifiable for speech watermarking. Table 3.1 shows the main differences between audio and speech watermarking requirements. Some parts of Table 3.1 are obtained from other studies [9, 13]. As a result, audio watermarking technology may not be a good candidate to apply for watermarking the speech signal due to the need to use more bandwidth and more amount of distortion.

3.3 Linear Predictive Analysis (LPA)

Speech is generated when air is exhaled from the lungs and passes through the throat, vocal cords, mouth, and nasal tract. This process can be modeled as in Eq. (3.1):

$$S(z) = G(z).H(z).R(z) \qquad (3.1)$$

where $S(z)$ corresponds to the original speech signal, $G(z)$ is modeled as impulse train for voiced or white noise for unvoiced, and $R(z)$ is modeled as in Eq. (3.2), while a fixed differentiator [14] and $H(z)$ are the vocal tract systems.

Table 3.1 Comparison between audio and speech watermarking

Criteria	Speech watermarking	Audio watermarking
Imperceptibility	Can be low	Should be high
Bandwidth	<16 kHz	>20 kHz
Channel noise	May be high	Must be low
Frequency spectrum	Randomly and noise-like	Tonal-like
Tonality	Voice tonality	Multiple tonality

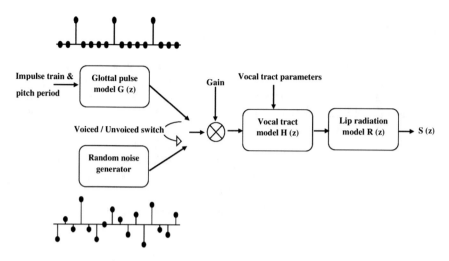

Fig. 3.1 Block diagram of speech production model [15]

$$R(z) = 1 - \alpha z^{-1} \ (0.9 < \alpha < 1) \tag{3.2}$$

where α is a constant $(0.9 < \alpha < 1)$, $Z = re^{\frac{i2\pi f}{Fs}}$ corresponds to polar number, r is its magnitude, $i2\pi f/Fs$ is its phase (angle), and Fs is the sampling frequency. Figure 3.1 shows the speech production model based on the source-tract filter model. As shown, the source excitation signal can be either periodic impulse (pitch) or white Gaussian noise. As depicted in Fig. 3.1, the source excitation and vocal tract can be separated from the speech signal by applying linear predictive analysis (LPA).

LPA can model quasi-stationary (between 20 and 30 ms) part of a speech signal as a linear combination of past samples. LPA models the vocal tract system with Pth-order (P) real linear predictive coefficients (LPCs) $[a_1, a_2, ..., a_p]$ as in Eq. (3.3), and LP residual error provides information about the excitation source as in Eq. (3.4):

$$\hat{S}(n) = \sum_{i=1}^{P} a_k s(n - k) \tag{3.3}$$

$$\text{LP residual error} = S(n) - \hat{S}(n) \tag{3.4}$$

Equation (3.5) explains how P is defined based on the sampling frequency (Fs):

$$P = \left(\frac{Fs}{1000}\right) + 2 \tag{3.5}$$

3.4 Speech Watermarking Techniques

Similar to audio watermarking techniques, various domains such as the time domain, transform domain, and compressed domain are applied for speech watermarking. Other techniques are available such as patchwork method which applies statistical methods to change the distance between two sets of the variance, the energy, or the mean of the signal for embedding the watermark data. However, the main techniques for speech watermarking can be classified into the following categories: auditory masking, phase modulation, quantization, transformation, and parametric modeling. Figure 3.2 shows an overview of the different attacks and techniques for speech watermarking.

In the remainder of this section, different speech watermarking techniques are presented in detail and the main advantages and disadvantages of each of them are discussed.

3.4.1 Speech Watermarking Based on Auditory Masking

In speech watermarking based on auditory masking techniques, some parts of the speech signal, which carries unimportant perceptual information for HAS, are watermarked to maximize the inaudibility of the watermarked data. To achieve this purpose, the maskee (lower sound which is not heard) with respect to masker (the louder sound) is detected based on spectral (frequency) domain or temporal (time) domain. Then, the maskee is watermarked to improve the imperceptibility of the watermarked speech signal [16, 17].

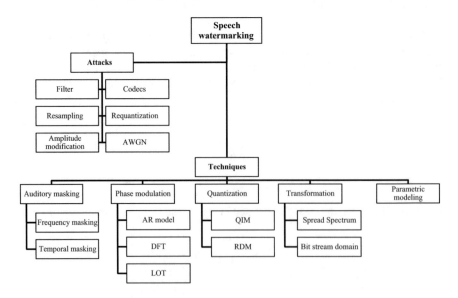

Fig. 3.2 An overview on speech watermarking

3.4.1.1 Frequency Masking

Basically in frequency domain, a strong signal makes mask a weaker signal when both of them have concurrent and close frequencies with respect to each other. There is a nonlinear hearing threshold model in frequency domain which can determine the minimum sound pressure level (SPL) for each frequency to be heard by HAS. This model reveals that the middle-frequency parts have more important perceptual information than high-frequency parts [18, 19].

Advantages

- The use of HAS in the frequency domain to prevent the listener to detect a faint tone for watermark embedding purpose.
- The embedding of the watermark in the inaudible regions of the speech signal to provide very high imperceptibility.

Disadvantages

- Due to the masking effects in high-frequency regions, embedding based on the HAS for speech is not efficient.
- High computational complexity.
- The watermark is embedded in high-frequency regions which cannot provide reasonable robustness.

3.4.1.2 Temporal Masking

There are two types of temporal masking in time domain: post-masking and pre-masking. In post-masking, a stronger masker produces a masking region immediately after 50–200 ms where another weaker maskee cannot be heard by HAS. However, finding a masking region in premasking is much more challengeable than post-masking [20, 21].

Advantages

- Use premasking and post-masking in the time domain to be inaudible for HAS.
- Easy to implement.

Disadvantages

- Premasking and post-masking require complicate computation to find the embedding regions.
- Low capacity and lack of transparency and security.

3.4.2 Speech Watermarking Based on Phase Modulation

Speech watermarking based on the phase modulation method embeds the watermark bits by modifying the phase of speech. The power spectrum is preserved without any change. Instead of MSE distortion as in other methods, in the phase modulation method, the watermarked speech and the original speech have the same power spectrum.

Generally, the phase of the speech strongly affects the representation of the speech signals. There is no standard definition available. There are three models for the speech phases, but none of them affects the speech spectral. These models are described as follows.

3.4.2.1 Autoregressive (AR) Phase Model

The autoregressive (AR) phase of the unvoiced speech signal is considered as a white Gaussian excitation. When this phase is manipulated, HAS cannot distinguish between the signals. This method does not affect the speech spectral and temporal envelope of the speech signal [22].

Advantages

- Low computational complexity due to the use of parametric modeling.
- High amount of capacity due to applying the watermark spectral shaping.
- This model does not need the original speech signal to extract the watermarks at the receiver side.
- High imperceptibility because the watermark signal is embedded in the phase of the non-voice part of the speech signal.

Disadvantages

- Low robustness against signal processing attacks.
- This model is so sensitive in comparison with attacks.
- Designing the watermark technique is difficult from one channel to another channel as the shaping of the watermark pulse is channel-dependent and designed for specific channel attack.

3.4.2.2 DFT Phase Model

By applying the discrete Fourier transform (DFT), the speech signal is modeled as complex coefficient and phase angle as in Eq. (3.6):

$$X = re^{j\phi} = a + jb$$
$$r = \sqrt{a^2 + b^2}, \ \phi = Arc\tan\left(\frac{b}{a}\right). \tag{3.6}$$

where b is imaginary coefficient, a is real coefficient, φ is phase, r is magnitude, and X is complex variable. Replacing the phase of the original speech signal with watermark's phase in DFT domain without changing the magnitude of the original speech signal has been developed by some studies [23, 24].

Advantages

- Basic and simple technique.
- This model embeds low distortion to the speech signal which gives better imperceptibility.

Disadvantages

- Low capacity because the watermarking embeds into just the DFT phase of the speech signal.
- Low robustness because the watermarking is not dispersed over the entire speech signal. Furthermore, phase coefficients are dependent which makes it difficult to extract the watermark bits accurately.

3.4.2.3 Lapped Orthogonal Transforms

Instead of DFT complex coefficient, the lapped transform coefficient is real and there is no phase available directly. The lapped transform works like filter banks where the variance of each short window sub-band is considered as the quantity that determines the spectral envelope and the particular realization of the sub-band signals as the phase of the signals. Therefore, replacing the original sub-bands with various sub-bands containing identical short-term envelope power can modify the speech phases [25].

The analysis by synthesis of the sampled orthogonal transform, which is very strong in the extraction procedure, is useful for speech watermarking. Two main lapped transforms, extended and modulated (ELT and MLT), are applied in phase modulation digital speech watermarking [26, 27].

Advantages

- This model introduces less distortion due to the way the spectral envelope of the host speech signal is preserved.
- The use of time–frequency transforms provides direct manipulation of the speech signal's phase.

Disadvantages

- Low imperceptibility due to the randomization of the perceptually coefficients.
- Low capacity due to the low number of sub-bands for watermarking.
- High computational complexity due to the use of frequency transformation.
- A trade-off between time and frequency which affects the robustness.

3.4.3 Speech Watermarking Based on Quantization

In contrast to phase modulation techniques and masking models, quantization techniques embed the watermark in the perceptually irrelevant segments of the speech signal [28]. The quantization-based speech watermarking techniques have improved the capacity due to the embedding of the watermark in perceptually irrelevant and relevant components of the speech signal. The technique can be divided into quantization index modulation (QIM) and rational dither modulation (RDM).

3.4.3.1 QIM

QIM is a more popular technique of modulation in watermarking scheme that uses Costa scheme [29]. Two main steps in QIM are index modulation and quantization. In index modulation step, an index or an indices' sequence is modulated with the inserted data. In the quantization step, the original signal is quantized by applying the associated quantizer or quantizers' sequence.

3.4.3.2 RDM

For coping with gain attacks on RDM, QIM can be extended [30]. RDM utilizes a proper function to embed the watermark bits in the ratio of the past watermarked samples and current sample nonlinearly.

Advantages

- Quantization is valuable in terms of the simple signal processing attacks such as linear filtering, nonlinear operations, AWGN, mixing, or resampling attack.
- Low computational complexity and easy to implement.
- The quantization step can be set independently without requiring the original speech signal (blindness).
- It is difficult to statistically analyze the watermark signal for steganalysis purposes.

Disadvantages

- Quantization is sensitive in terms of amplitude scaling attack.
- Low capacity as only one watermark bit is embedded by using quantization operation.
- When the speech signal is heavily quantized, the imperceptibility of the speech signal is degraded.

3.4.4 Speech Watermarking Based on Transformation

Two major speech watermarking techniques are available based on the transformation, which are described in the following:

3.4.4.1 Spread Spectrum

Broadening a narrowband signal with broadband carrier signal in frequency domain by applying modulation technique is known as spread spectrum. For this purpose, a pseudo-noise (PN) sequence is applied on the watermark signal to spread its spectrum in frequency domain. As watermark signal is spread over a larger range of spectrum and is spread over a larger set of samples, impairment and removal attacks cannot degrade spread-spectrum speech watermarking technique. The robustness can be improved when important perceptual part of the speech spectrum is applied for watermarking [31, 32].

Advantages

- Provides good robustness.
- Difficult to remove the embedded watermark.

Disadvantages

- Needs to apply decomposition function (such as DFT, DCT, and DWT) as well as inverse decomposition function to convert the signal. This technique requires more time and delay.
- Spread spectrum is not good candidate for a real-time watermarking system.

3.4.4.2 Bitstream Domain

The bitstream domain embeds the watermarked data directly during or after compression [33, 34]. Embedding the watermark data into relevant parts of the original speech signal during or after compression can improve the robustness of this technique against the compression attacks. These watermarking techniques can embed the watermarked bits into the bitstream of the codec, e.g., G.723.1, G.711-PCMU, G.729, and ACELP, to bypass the compression attacks [35, 36].

Advantages

- Provides better robustness in terms of compression attacks.
- Low computational complexity.
- Supports all bit rate.
- Effective for real-time watermarking applications.

Disadvantages

- Very low embedding capacity.
- Not robust in terms of some attacks such as D/A conversion.

3.4.5 Speech Watermarking Based on Parametric Modeling

Although there is no generation model available for the image or audio signal, the digital speech signal can be modeled (Sect. 3.3). Parametric modeling technique utilizes LPCs or line spectrum pair (LSP) or log area ratio (LAR) indirectly to modify or quantize (AR) the parameters for embedding the watermark. These coefficients are not quantized directly because they cannot be controlled by the pole location of AR model. Therefore, the system's stability cannot be guaranteed [37, 38].

Advantages

- Effective technique which is always one part of speech processing application such as speaker and speech recognition, speech coding, speech synthesis, and speech watermarking.
- Low computational complexity as computing the LPCs requires simple autocorrelation.
- Ability to integrate with the bitstream domain watermarking technique.
- Saves data storage space.

Disadvantages

- Even without any attack, the embedded and extracted LPCs are different from one another.
- Very low robustness.
- Difficult to keep the stability of the LPCs.
- Low capacity.

3.5 Attacks for Speech Watermarking

Figure 3.2 shows different types of speech watermarking attacks. Although some of the attacks have already discussed in Chap. 2, the major speech watermarking attacks are discussed in this section. These attacks are mainly taking place through speech telephony channels.

3.5.1 Additive White Gaussian Noise (AWGN)

This attack simulates the distortion in telephony or network channels. This attack adds some uniform frequency values with Gaussian distribution to the watermarked speech signal. This attack can change the value of SNR of the watermarked signal. Whenever the SNR value is decreased by AWGN, the robustness of the speech watermarking is decreased.

3.5.2 Low-Pass Filter (LPF)

In some telephony channel, low-pass filter is applied to attenuate frequencies higher than a specific value which is called cutoff frequency. This attack can degrade the robustness of a watermarked signal if watermarked is embedded into the higher frequencies of the speech signal.

3.5.3 Band-Pass Filter (BPF)

Similar to LPF, a BPF is performed on the watermarked signal to pass certain frequencies between two specific frequencies. For example, the cutoff frequencies for G.711 are between 300 and 3400 Hz. Therefore, a BPF is required to simulate the G.711 VoIP channel. Figure 3.3 shows the frequency range for narrowband G.711 and wideband G.722. Basically, from 64 kbps of voice stream in PSTN telephony channel ($R = $ sample rate $\times$ bit depth $\times$ channel $= 8000 \times 8 \times 1 = 64$ kbps), G.722 applies adaptive differential pulse-code modulation (ADPCM) to allocate 48 kbps for low frequencies (0–4 kHz) and 16 kbps for high frequencies (4–8 kHz). Basically, G.722 is selected due to its generality, integrity, and popularity. Figure 3.3 illustrates the frequency ranges for both G.711 and G.722 VoIP channels.

3.5.4 A-Law

An A-law is an algorithm for companding the speech signal based on 8-bit PCM which is generally used in Europe. A-law is one of the versions for G.711 presented by ITU-T. The prevailing parameter for A-law is usually considered as $A = 87.6$.

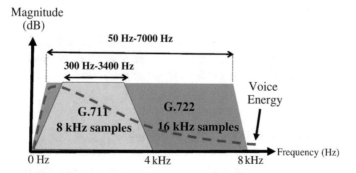

Fig. 3.3 Frequency range of G.711 (narrowband) and G.722 (broadband) with voice energy concentration slope in frequency domain

3.5.5 μ-Law

Similar to A-law, μ-law is an algorithm for companding the speech signal based on 8-bit PCM which is generally used in Japan and North America. μ-law is one of the versions for G.711 presented by ITU-T. The prevailing parameter for A-law is usually considered as $\mu = 255$.

3.5.6 CELP

CELP is a speech coding technique that applies analysis-by-synthesis (AbS) approach based on LPA to code a speech signal. MPEG-4 Audio is using CELP for compression.

3.5.7 Amplitude Variation

Sometimes, amplitude of a speech signal is increased or decreased by multiplying and dividing via a specific value, respectively. This gain value can degrade the robustness of the speech watermark.

3.5.8 Resample

In order to change the sampling frequency and bandwidth of a signal, resampling technique is applied to convert one sampling rate to another sampling rate. This rate can be upsampled or downsampled.

3.5.9 Requantization

Quantization is a process in digitalization of a signal that limiting large continuous set to small discrete set. Requantization is quantized the watermarked signal to specific bit rate and then are requantized to the original one.

3.6 Challenges in Speech Watermarking

Speech watermarking is a trade-off between robustness, impermeability, and capacity. Depending on the applications, some techniques focus on robustness. However, other applications emphasize imperceptibility and inaudibility. In many applications,

Table 3.2 Comparison of related audio and speech watermarking methods

WM method	Capacity (bps)	Robustness (BER %)	Imperceptibility (SNR)	Fs (Hz)
Speech watermarking methods				
Frequency masking [39]	146	0.001	25	4000
Temporal masking [21]	352	24.86	40.4	44,100
AR model [40]	243	2	5	7000
AR model [10]	690	6.6	30	27,000
QIM of DFT [41]	40	7.5	11	8000
Bitstream domain [33]	1600	N/A	20.3	N/A
Spread spectrum [31]	24	5	20	28,000
Spread spectrum [32]	800	28.4	∞	4000
Parametric modeling [38]	N/A	2	33.4	8000
Parametric modeling [42]	4	3	∞	6000
Alternative audio watermarking methods				
LOT-RDM [43]	689	0	63.2	44,100
SVD-DWT [44]	258	0.61	38.17	44,100
Adaptive DWT-SVD [45]	45.9	13.77	24.37	44,100
DWT based [46]	28.71	1.8	23.98	44,100
SVD-STFT based [47]	32	0.13	N/A	44,100
QIM of DCT of DWT [48]	420	0	10	22,500
LWT and SVD [49]	170.67	14.18	22.025	44,100
SVD-DCT [50]	43	0.19	29.679	44,100
DWPT-SVD-adaptive QIM [51]	139.97	0.43	20.327	44,100

the perceptually irrelevant segments of the speech are attempted to be removed. This removal is a basic challenge for speech watermarking algorithm so as to preserve its robustness. Watermarking should occur perceptually in every relevant speech part. However, the limitation on the number of perceptually relevant segments is another concern for watermarking. In some applications, the capacity becomes more important to achieve. Furthermore, the capacity of speech watermarking is reduced as compared to that of audio due to the narrow usable bandwidth. This embedding capacity can be further reduced, when speech codec techniques are applied. Table 3.2 summarizes the evaluation of the performance of each watermarking technique in terms of robustness, imperceptibility, and capacity. As shown, none of them is ideal due to the nature of watermarking which is a trade-off among these criteria.

References

1. Nematollahi, M.A., and S. Al-Haddad. 2015. Distant speaker recognition: An overview. *International Journal of Humanoid Robotics* 1550032.
2. William, S. 2006. *Cryptography and network security*, 4 edn. Pearson Education India.

3. Huang, H.-C., and W.-C. Fang. 2010. Metadata-based image watermarking for copyright protection. *Simulation Modelling Practice and Theory* 18(4): 436–445.
4. Huang, H.-C., et al. 2011. Tabu search based multi-watermarks embedding algorithm with multiple description coding. *Information Sciences* 181(16): 3379–3396.
5. Faundez-Zanuy, M., J.J. Lucena-Molina, and M. Hagmüller. 2010. Speech watermarking: An approach for the forensic analysis of digital telephonic recordings*. *Journal of Forensic Sciences* 55(4): 1080–1087.
6. Faundez-Zanuy, M. 2010. Digital watermarking: New speech and image applications. *Advances in Nonlinear Speech Processing,* 84–89.
7. Faundez-Zanuy, M., M. Hagmüller, and G. Kubin. 2006. Speaker verification security improvement by means of speech watermarking. *Speech Communication* 48(12): 1608–1619.
8. Faundez-Zanuy, M., M. Hagmüller, and G. Kubin. 2007. Speaker identification security improvement by means of speech watermarking. *Pattern Recognition* 40(11): 3027–3034.
9. Hagmüller, M., et al. 2004. Speech watermarking for air traffic control. *Watermark* 8(9): 10.
10. Hofbauer, K., G. Kubin, and W.B. Kleijn. 2009. Speech watermarking for analog flat-fading bandpass channels. *IEEE Transactions on Audio, Speech, and Language Processing* 17(8): 1624–1637.
11. Hofbauer, K., H. Hering, and G. Kubin. 2005. Speech watermarking for the VHF radio channel. In *Proceedings of the 4th Eurocontrol innovative research workshop*.
12. Rabiner, L.R., and R.W. Schafer. 1978. *Digital processing of speech signals*. Prentice Hall.
13. Al-Shoshan, A.I. 2006. Speech and music classification and separation: A review. *Journal of King Saud University* 19(1): 95–133.
14. Flanagan, J.L. 1972. *Speech analysis: Synthesis and perception*.
15. Rabiner, L.R., and R.W. Schafer. 2009. *Theory and application of digital speech processing*. Preliminary Edition.
16. Blamey, P., et al. 1987. Acoustic parameters measured by a formant-estimating speech processor for a multiple-channel cochlear implant. *The Journal of the Acoustical Society of America* 82(1): 38–47.
17. Schroeder, M.R., B.S. Atal, and J. Hall. 1979. Optimizing digital speech coders by exploiting masking properties of the human ear. *The Journal of the Acoustical Society of America* 66(6): 1647–1652.
18. Taal, C.H., R.C. Hendriks, and R. Heusdens. 2012. A low-complexity spectro-temporal distortion measure for audio processing applications. *IEEE Transactions on Audio, Speech, and Language Processing* 20(5): 1553–1564.
19. Swanson, M.D., et al. 1998. Robust audio watermarking using perceptual masking. *Signal Processing* 66(3): 337–355.
20. Bassia, P., I. Pitas, and N. Nikolaidis. 2001. Robust audio watermarking in the time domain. *IEEE Transactions on Multimedia* 3(2): 232–241.
21. Cvejic, N., A. Keskinarkaus, and T. Seppanen. 2001. Audio watermarking using m-sequences and temporal masking. In *IEEE workshop on the applications of signal processing to audio and acoustics, 2001*. IEEE.
22. Kubin, G., B. Atal, and W. Kleijn. 1993. Performance of noise excitation for unvoiced speech. In *Proceedings of IEEE workshop on speech coding for telecommunications, 1993*. IEEE.
23. Kim, D.-S. 2003. Perceptual phase quantization of speech. *IEEE Transactions on Speech and Audio Processing* 11(4): 355–364.
24. Takahashi, A., R. Nishimura, and Y. Suzuki. 2005. Multiple watermarks for stereo audio signals using phase-modulation techniques. *IEEE Transactions on Signal Processing* 53(2): 806–815.
25. Malvar, H.S. 1992. *Signal processing with lapped transforms*. Artech House.
26. Malvar, H.S. 1992. Extended lapped transforms: Properties, applications, and fast algorithms. *IEEE Transactions on Signal Processing* 40(11): 2703–2714.
27. Shlien, S. 1997. The modulated lapped transform, its time-varying forms, and its applications to audio coding standards. *IEEE Transactions on Speech and Audio Processing* 5(4): 359–366.
28. Cox, I.J., et al. 2002. *Digital watermarking*. Vol. 53. Springer.

29. Costa, M.H. 1983. Writing on dirty paper (corresp.). *IEEE Transactions on Information Theory* 29(3): 439–441.
30. Chu, W.C. 2004. *Speech coding algorithms: Foundation and evolution of standardized coders*. Wiley.
31. Arora, S. and S. Emmanuel. 2003. Adaptive spread spectrum based watermarking of speech. In *9th National undergraduate research opportunities programme congress* Poster 15.
32. Cheng, Q. and J. Sorensen. 2001. Spread spectrum signaling for speech watermarking. In *Proceedings (ICASSP'01) IEEE international conference on acoustics, speech, and signal processing, 2001*. IEEE.
33. Geiser, B. and P. Vary. 2008. High rate data hiding in ACELP speech codecs. In *IEEE international conference on acoustics, speech and signal processing, 2008. ICASSP 2008*. IEEE.
34. Lacy, J., et al. 1998. On combining watermarking with perceptual coding. In *Proceedings of the 1998 IEEE international conference on acoustics, speech and signal processing, 1998*. IEEE.
35. Liu, C.-H. and O.T.-C. Chen. 2004. Fragile speech watermarking scheme with recovering speech contents. In *The 2004 47th midwest symposium on circuits and systems, 2004. MWSCAS'04*. IEEE.
36. Zhe-Ming, L., Y. Bin, and S. Sheng-He. 2005. Watermarking combined with CELP speech coding for authentication. *IEICE Transactions On Information And Systems* 88(2): 330–334.
37. Yan, B., and Y.-J. Guo. 2013. Speech authentication by semi-fragile speech watermarking utilizing analysis by synthesis and spectral distortion optimization. *Multimedia Tools And Applications* 67(2): 383–405.
38. Gurijala, A. 2007. *Speech watermarking through parametric modeling*. ProQuest.
39. Chen, S. and H. Leung. 2006. Concurrent data transmission through PSTN by CDMA. In *Proceedings of 2006 IEEE international symposium on circuits and systems, 2006. ISCAS 2006*. IEEE.
40. Malik, H.M., R. Ansari, and A.A. Khokhar. 2007. Robust data hiding in audio using allpass filters. *IEEE Transactions on Audio, Speech, and Language Processing* 15(4): 1296–1304.
41. Narimannejad, M. and S.M. Ahadi. 2011. Watermarking of speech signal through phase quantization of sinusoidal model. In *19th Iranian conference on electrical engineering (ICEE), 2011*. IEEE.
42. Hatada, M., et al. 2002. Digital watermarking based on process of speech production. In *ITCom 2002: the convergence of information technologies and communications*. International Society for Optics and Photonics.
43. Garcia-Hernandez, J.J., M. Nakano-Miyatake, and H. Perez-Meana. 2008. Data hiding in audio signal using rational dither modulation. *IEICE Electronics Express* 5(7): 217–222.
44. Al-Haj, A. 2014. An imperceptible and robust audio watermarking algorithm. *EURASIP Journal on Audio, Speech, and Music Processing* 2014(1): 1–12.
45. Bhat, V., I. Sengupta, and A. Das. 2010. An adaptive audio watermarking based on the singular value decomposition in the wavelet domain. *Digital Signal Processing* 20(6): 1547–1558.
46. Xiang, S. 2011. Audio watermarking robust against D/A and A/D conversions. *EURASIP Journal on Advances In Signal Processing* 2011: 3.
47. Özer, H., B. Sankur, and N. Memon. 2005. An SVD-based audio watermarking technique. In *Proceedings of the 7th workshop on multimedia and security*. ACM.
48. Wang, X., W. Qi, and P. Niu. 2007. A new adaptive digital audio watermarking based on support vector regression. *IEEE Transactions on Audio, Speech, and Language Processing* 15(8): 2270–2277.
49. Lei, B., et al. 2012. A robust audio watermarking scheme based on lifting wavelet transform and singular value decomposition. *Signal Processing* 92(9): 1985–2001.
50. Lei, B.Y., I.Y. Soon, and Z. Li. 2011. Blind and robust audio watermarking scheme based on SVD–DCT. *Signal Processing* 91(8): 1973–1984.
51. Hu, H.-T., et al. 2014. Incorporation of perceptually adaptive QIM with singular value decomposition for blind audio watermarking. *EURASIP Journal on Advances in Signal Processing* 2014(1): 1–12.

Part II
Multimedia Watermarking

Chapter 4
Image Watermarking

4.1 Introduction

Visual system is the most critical sense for human due to bringing memories, learning, and interfering about an event. An image can reveal more information about an event which even one hundred sentences cannot explain. Therefore, images and pictures are becoming one of the most conventional parts of human life worldwide. This is due to the ability of imaging devices which facilitate easy imaging skill for people to record every moment of their life. However, the popularity of these images makes some issues such as ownership, authentication, and security. Digital watermarking as a potential technology tries to rectify these issues by embedding extra information into the original image in an invisible way.

This chapter provides information about image watermarking and some preliminary information about image processing which have major contributions in secure systems. Furthermore, this chapter reviews traditional approaches and related works on image watermarking to reveal the advantages and disadvantages of each technique.

4.2 Background

A grayscale image consists of some integer values between 0 and 255 in two-dimensional arrangements. On the other hand, a color image consists of three channels including red channel, green channel, and blue channel, in which each of them can be considered as a grayscale image. Currently, image processing has the most important contribution in computer science which requires a strong security mechanism such as digital watermarking to cover security issues in all image applications.

© Springer Science+Business Media Singapore 2017
M.A. Nematollahi et al., *Digital Watermarking*, Springer Topics
in Signal Processing 11, DOI 10.1007/978-981-10-2095-7_4

Generally, image watermarking has wide applications (law enforcement application and commercial application) such as copy control, security issues, surveillance, and bankcard identification. The main application is for copyright. In this application, the watermark is embedded in image to protect the ownership.

4.3 Image Watermarking Techniques

Various domains including the spatial domain, transform domain, and multiple domains are applied for digital image watermarking. Other techniques are available such as multiplication method which applies statistical methods to change the distance between two sets of the variance, the energy, or the mean of the signal for embedding the watermark data. However, the main techniques for digital image watermarking can be classified into the following categories: spatial domain, transform domain, and multiple domain. Figure 4.1 shows an overview of the different categories and subcategories for each digital image watermarking technique:

4.3.1 Spatial Domain

A few systems have been formed to watermark embedding in spatial space. Additive, LSB replacements, and singular value decomposition (SVD) are well-known techniques for image watermarking in spatial domain. These methodologies utilize straightforward manipulation on the pixel qualities of the original image.

In this way, a pseudorandom number sequence (pattern noise) is generated and then embedded directly in image pixel intensity. This pattern usually has an integer value such as $(-1, 0, 1)$ or sometimes float value. For ensuring that watermark can be extracted, using an established pattern noise is a key that the correlation between the numbers of keys will be a little different. Embedded binary sequence $W \in \{0, 1\}$ in original image can show picture using Eq. (4.1) [1]:

$$\text{Watermarked image} = \text{Original Image} + \alpha \times W \qquad (4.1)$$

Where α is the intensity of the watermark. For watermark detection which is non-blind, the cross-correlation between W and watermarked image, which is passed or attacked, has computed.

In LSB technique, less relevant maintenance information of an image is modified which does not affect the perceptibility of an image. The basic idea here is the LSB of an 8-bit grayscale image can carry the watermark bit without significant modification of the original byte.

Instead of frequency transformation, there are strong mathematical decomposition which can be efficient than the basic frequency transformation techniques. SVD is one of these strong numerical techniques, which is widely applied in

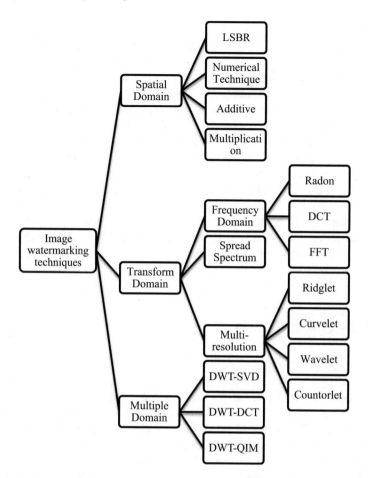

Fig. 4.1 Image watermarking techniques

image watermarking techniques. SVD can decompose a rectangular matrixes into three basic matrixes such as left eigenvector, eigenvalue, and right eigenvector as in Eq. (4.2). The eigenvalue matrix is a diagonal matrix in which the eigenvalues are ordered as descending order like $\sigma(i) > \sigma(i+1)$.

$$A = USV^T = [U11, \ldots, U_{nm}] \times [S11, \ldots, S_{nn}] \times [V11, \ldots, V_{rn}]^T \quad (4.2)$$

Basically, small manipulation of the eigenvalues cannot seriously affect the quality of the image signal. Therefore, adding the watermark in eigenvalues can be robust and most possible approach. However, providing enough trade-off among capacity, invisibility, and robustness for this technique can be the challenging problem.

Multiplication watermarking techniques are widely used for media watermarking. Basically, multiplication can be embedded in both special domain and transformed domains. Also, it is very common for image watermarking due to simplicity during embedding and extraction. A few approaches have been developed for image based on the multiplication. This is due to mathematical difficulty in threshold for watermark extraction. In [2], image watermarking technique has been developed based on scaling. The watermark is adapted somehow it cannot see by HVS. For extraction, maximum likelihood (ML) technique is applied which is mathematically computed. Not only this technique can provide robustness, but also it can provide enough imperceptibility. Another approach has been developed based on statistical modeling of multiplication a constant in contourlet transform of the image [3]. Since the watermark is embedded in edge of the image, watermark is embedded by more robustness which can provide less BER with respect to AWGN, JPEG, and filtering attacks. Furthermore, ML optimum detector and decision threshold are mathematically developed in order to reduce the BER at the receiver side. In [4], image watermarking technique is developed based on the statistical modeling of the image histogram mean and shape. This technique has provided good robustness against translation, scaling, cropping, random bending, and rotation attacks.

4.3.2 Transform Domain

Currently, there are many attentions in applying transformation domain for image watermarking. Instead of modification of pixel value in special domain, transform domain has modified the transform value of the images. There are many transformation tools which are widely based on the frequency transformations such as DFT, DCT, and wavelet transforms. Although watermarking in the transform domain is more stronger than the special domain, huge amount of complexity, time, and memory are the main disadvantages of the transform domain. For example, DCT needs a huge amount of computational complexity to transform an image to DCT domain. Therefore, the majority of digital image watermarking approaches have only applied a block with small size, i.e., 8 × 8, to consume computer resources. Also, fast Fourier transform (FFT) is the efficient implementation of DFT.

In [5], a comparison has been made to evaluate the robustness of the spatial and transform domains. It is shown that the robustness of the watermark is increased when it is embedded in transform domain. Furthermore, embedding the watermark in frequency domain can seriously improve the extraction process under geometrical and non-geometrical image's attacks. In [6], wavelet transformation has been done to computer different sub-bands in digital image. Then, the watermark bits are embedded by applying quantization index modulation (QIM) which not only provides blindness but also shows good robustness in terms of various types of attacks. However, it cannot provide good capacity for embedding a biometric watermark in image.

In [3], an image watermarking is proposed based on the contourlet transformation. By using contourlet transform, the visibility of the watermarked image has been improved because HVS is less sensitive to edge where watermark is embedded in this technique. In [7], two main researches on watermarking have been done to embed the watermark in magnitude and phase of Fourier domain, respectively.

Another digital image watermarking is proposed based on the redundant discrete wavelet transform (RDWT) [8]. This technique applied red and blue components of color image in order to reduce the visibility of the watermark in watermarked image. In this technique, the watermark bits are inserted semi-blindly after RDWT is applied on image. Although it seems that this technique has fair capacity due to embedding MFFC feature, the robustness of this technique is not reasonable which can be robust in real-channel transformation.

Generally, transform domain has some advantages as compared to spatial domain. First, it can be more robust due to embedding the watermark in low-frequency region. Second, it is more robust under different attacks. Third, the amount of injected dissertation can be controlled which is more difficult to achieve for spatial domain.

4.3.3 Multiple Domains

In this part, multiple transformations have been reviewed. In this technique, the image watermarking has been developed based on the combination of different transformations in the direction of providing more robustness. There are different methodologies which are fully described here. A blind DWT-DFT image watermarking technique has been proposed which can provide enough robust against JPEG compression and affine transformation, but in the lack of robustness against filtering, median is the main limitation of this approach. Another approach is proposed based on the DCT-DWT for image authentication where compression is taken place. This image authentication has applied temper detection in special part of the image which can provide soft authenticator as well as chrominance. The other watermarking technique is developed by the combination of DCT and SVD. In this approach, local peak signal-to-noise ratio (LPSNR) has been applied to make the watermark robust against Stir Mark 4.0 attack.

Same DCT and SVD image watermarking techniques are developed in order to improve the robustness against different signal processing manipulation such as noise, JPEG compression, median filter, low-pass filter, and contrast enhancement. Recently, combination of DWT and SVD is also proposed for authenticity and copyright protection of digital images. The main advantage of this method is that information cannot be extracted without the original image. Also, LWT-SVD robust image watermarking approach has been developed. Although this approach can provide enough robustness against geometrical attacks, it cannot provide enough invisibility.

4.4 Image Attacks

Basically, many intentional and unintentional attacks may perform on water-marked image. In order to design robust and fragile image watermarking techniques, the most common attack for image watermarking should be known. These attacks are common during the image transmission, compression, and manipulations as shown in Fig. 4.2.

Removal attacks are kinds of attacks that aim to evacuate the watermark data from the watermarked image without endeavoring should break those security of the watermarking calculation. This kind of watermark ambush does not endeavor and will figure out those encryption strategies utilized or how the watermark need

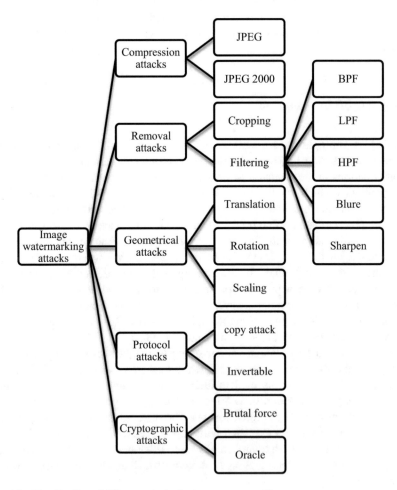

Fig. 4.2 Classification of different attacks for image watermarking

has been installed. It brings about a degraded watermarked image, consequently a degraded watermark signal, the place no basic post preparing could recuperate those watermark data from the assaulted information. Incorporated in this class are noising, histogram equalization, smudge, and hone attack. Incorporated in this class are noising, histogram equalization, smudge, and hone attack.

Geometry attack would rather unique in relation to evacuation strike. As opposed to pointing to uproot or extremely harming those watermark signals, this sort from claiming attack means on misshaping the watermark indicator. It still can be considered as a moral attack if the point of interest of the geometry assault might make secured. Also, a countermeasure is connected. Those methodologies from claiming correcting this sort of ambush will be often alluded with similarly as synchronization. However, those unpredictability of the required synchronization methodology could be considered as awful prohibitively exorbitant. Incorporated in this class about watermark strike are picture rotation, scaling, interpretation what's more skewing.

The point from claiming cryptographic attack should split those security systems clinched alongside watermarking schemes and subsequently find an approach with uproot those inserted watermark majority of the data or on implant misdirecting watermarks. A standout among the strategies in this classification may be those brute-force quest strategies. This system extensively endeavors should split the watermark security. However, Tom eventually was utilizing an expansive scan for the serious mystery data. On the other hand, Prophet attack is using non-watermarked content at watermark extraction module for accessing.

Protocol attack is a further sort about watermark ambush. Inasmuch as the opposite sorts from claiming attack point during destroying, distorting alternately extracting the watermark signal, protocol attack includes those attacker's own watermark signs onto those information being referred to. This brings about ambiguities on the valid proprietorship of the information being referred to. Protocol attack targets the whole idea of utilizing watermarking systems similarly as an answer will copyright insurance.

Another protocol attack is the duplicate attack: As opposed to destroying the watermark, the duplicate attack estimates a watermark from watermarked information and also duplicates it on some other data, known as those focus information. Those assessed watermarks are adjusted of the nearby characteristics of the focus information on fulfilling its intangibility.

Compression attacks: While image/video is distributed to the public, it is very common to compress it or change the format. Different lossless and lossy techniques such as LZW, JPEG, JPEG 2000, and MPEG have been applied on image in order to reduce the size. Therefore, the watermark could be enough robustness against common compression attacks.

4.5 Challenges in Image Watermarking

Similar to audio/speech watermarking, the main challenge in image watermarking is trading off among capacity, invisibility (imperceptibility), and robustness which is application-dependent. Although there is possibility to embed the watermark in irrelevant visual parts of the image to improve the invisibility of the watermark, the robustness of the watermark is decreased due to the possibility of removing these irrelevant parts. Thus, watermarking uses perceptual image parts for embedding. Another limitation is related to watermark capacity which is directly depended on the size of the image and number of visually relevant part inside the image. In contrast to audio/speech signal, image has less number of samples; i.e., 1 s of audio has 44,100 samples, but an image with size of 100×100 has 10,000 samples. Furthermore, some limitations in forensic image watermarking are making challenges for the application of image watermarking widely. For instants, text image (such as bills, receipt, and contracts) as an electronic evidence has less embedding capacity which requires text watermarking or natural language watermarking to solve it. In addition, image authentication based on image watermarking requires embedding the watermark in image-forming process step which is a challenge since many cameras in the market do not have this ability. Although robust image watermarking and fragile image watermarking can solve ownership and authenticity problems in digital image forensic, respectively, judging the source and authenticity of an image simultaneously with image watermarking is a critical challenge for image watermarking.

4.6 Image Watermarking Based on Human Visual System

HVS may apply for image watermarking technology to increase the invisibility of the watermark. A visual signal must contain a minimum amount of contrast to be perceived by HVS. In addition, a given frequency for a signal should mask other disturbing similar frequencies to be seen by HVS [5, 9]. Thus, the contrast itself is depending on mean frequency and luminance. The variation in regions of high luminance is less sensitive to HVS, which makes the watermark to be a gain factor luminance dependent [10]. For color image, HVS is less sensitive to blue channel, red channel, and green channel in ascending order. Therefore, some studies have embedded the watermark in blue channel to enhance the watermark invisibility [10]. In contrast to the smooth area in an image, HVS is sensitive to edges and to textured areas of an image. As a result, some works have been used spatial masking effect to embed the watermark in smooth areas of an image [11]. In this line, some local segments with less notice for human eye are detected for increasing their energy. Whenever this energy is increased, a robust watermark can be embedded. As a result, image watermarking based on HVS is not simply distributing watermark over all parts of the image, but it is applying perceptual irrelevant parts of the image for embedding watermark with higher energy to guarantee the invisibility.

4.7 Medical Image Watermarking

In recent decade, e-health service is becoming one part of the modern medical system that allows transmitting and receiving remote medical image over electronic networks for improving various purposes such as clinical interpretation, standards, delivery, access, and health care. However, some security issues such as malpractice liability, image retention, fraud, and privacy are rising due to this remote medical image access. Although different international and national directives and legislative rules have defined privacy and security requirements for medical image, it seems these requirements are incapable to provide enough privacy and security. There are two main requirements which can be summarized in the following: first, definition of a standard set for image privacy and security; second, definition of a set of security measurements that fulfilled the security principles in the profile [12]. Therefore, image watermarking as attractive tool has enough properties that can complete these security requirements. Some justifications for applying image watermarking for medical image have been explained in [12, 13].

References

1. Zeng, W., and B. Liu. 1999. A statistical watermark detection technique without using original images for resolving rightful ownerships of digital images. *IEEE Transactions on Image Processing* 8(11): 1534–1548.
2. Akhaee, M.A., et al. 2009. Robust scaling-based image watermarking using maximum-likelihood decoder with optimum strength factor. *IEEE Transactions on Multimedia* 11(5): 822–833.
3. Akhaee, M.A., S. Sahraeian, and F. Marvasti. 2010. Contourlet-based image watermarking using optimum detector in a noisy environment. *IEEE Transactions on Image Processing* 19(4): 967–980.
4. Xiang, S., H.J. Kim, and J. Huang. 2008. Invariant image watermarking based on statistical features in the low-frequency domain. *IEEE Transactions on Circuits and Systems for Video Technology* 18(6): 777–790.
5. Ghazy, R.A., et al. 2015. Block-based SVD image watermarking in spatial and transform domains. *International Journal of Electronics* 102(7): 1091–1113.
6. Yassin, N.I., N.M. Salem, and M.I. El Adawy. 2014. QIM blind video watermarking scheme based on wavelet transform and principal component analysis. *Alexandria Engineering Journal* 53(4): 833–842.
7. Khan, M.K., L. Xie, and J. Zhang. 2007. Robust hiding of fingerprint-biometric data into audio signals. In *Advances in biometrics*, 702–712. Springer.
8. Vatsa, M., R. Singh, and A. Noore. 2009. Feature based RDWT watermarking for multimodal biometric system. *Image and Vision Computing* 27(3): 293–304.

9. Gopalakrishnan, K., N. Memon, and P.L. Vora. 2001. Protocols for watermark verification. *IEEE Multimedia* 8(4): 66–70.
10. Huang, Y.F., S. Tang, and J. Yuan. 2011. Steganography in inactive frames of VoIP streams encoded by source codec. *IEEE Transactions on Information Forensics and Security* 6(2): 296–306.
11. Swanson, M.D., et al. 1998. Robust audio watermarking using perceptual masking. *Signal Processing* 66(3): 337–355.
12. Coatrieux, G., et al. 2000. Relevance of watermarking in medical imaging. In *Proceedings of 2000 IEEE EMBS international conference on information technology applications in biomedicine, 2000*. IEEE.
13. Nyeem, H., W. Boles, and C. Boyd. 2013. A review of medical image watermarking requirements for teleradiology. *Journal of Digital Imaging* 26(2): 326–343.

Chapter 5
Video Watermarking

5.1 Introduction

Recently, video is the major part in the applications such as Internet multimedia, videophone, video on demand, set-top box, and video conference, which requires a security mechanism to protect it from intentional and unintentional attacks. Video watermarking can serve as an urgent secure component to enhance security and privacy of these applications.

Basically, video watermarking is just the enhanced version of image watermarking. Digital video is a sequence of consecutive still images known as frames. Unlike image watermarking, video watermarking also addresses the problem of large volume. In addition, video watermarking must overcome to extra attacks including frame swapping, frame averaging, video compression, and video coding. Providing balance between motionless and motion regions is also another concern in video watermarking. Generally, video watermarking algorithms hide the watermark directly to the video or to the transformed version of host video.

5.2 Background of Video

In digital video, three major pictures/frame types are used including I, P, and B. In the following, the characteristics of each I, P, and B are explained:

- I–frames (Intra-coded picture): This frame is the primary frame which is compressed minimally as compare to P and B frames. In addition, I-frame is not needed for P and B frames for decoding.
- P–frames (Predicted picture): This frame is decompressed based on previous frames and more than I-frame.

© Springer Science+Business Media Singapore 2017
M.A. Nematollahi et al., *Digital Watermarking*, Springer Topics
in Signal Processing 11, DOI 10.1007/978-981-10-2095-7_5

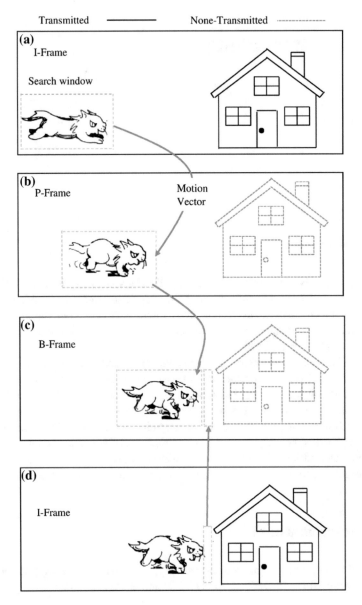

Transmitted ———— None-Transmitted ············

(a)
  I-Frame

Search window

(b)
  P-Frame   Motion
                                   Vector

(c)
  B-Frame

(d)
  I-Frame

Fig. 5.1 The sequence of video frames including. **a** I-frame. **b** P-frame. **c** B-frame. **d** I-frame, and other video components

- B–frames (Bipredictive picture): This frame is compressed more than I-frame and P-frame, and it is decompressed based on both forward and previous frames.

Figure 5.1 shows an example of how I, P, and B frames are constructing a video sequence. As seen, majority of the data in a new frame can be found in previous

frames but in different location. Thus, motion compensation (motion information) vector is used to code the video information with minimum amount of bits.
For this purpose, a video frame is divided into macroblocks which consists of
blocks of pixels. Then, each block can be compressed (coded) or decompressed
(predicted) based on the matching block in the reference frame. If the block is
matched between them, the position of the match in the reference frame is coded.

5.3   Video Watermarking Techniques

Lots of techniques have been presented in the past for video watermarking.
Techniques used in image watermarking can be applied directly to the video watermarking; therefore, many techniques of image watermarking are also used in video
watermarking. Every technique has its advantages and disadvantages. In video
watermarking, watermark is generally embedded in uncompressed video or some
time in compressed video [1]. In broader sense or in the basis of domain on which
watermark is inserted, video watermarking can be grouped into two different categories including video-based and image-based techniques as shown in the Fig. 5.2.

5.3.1   Image-Based Video Watermarking

As discussed, a video consists of some sequence of images. As a result, there is
possibility to apply image watermarking techniques for watermarking a video

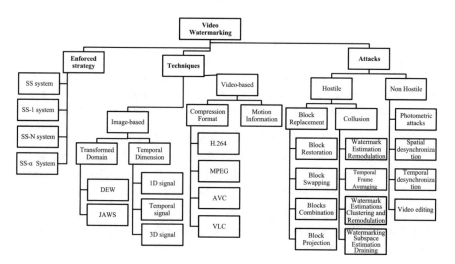

Fig. 5.2   Video watermarking an overview

content based on frame-by-frame approach. Similar to moving JPEG (M-JPEG) that compresses each frame of a video based on JPEG image compression technique, some video watermarking techniques have embedded data in each of a video frame [2]. In these watermarking techniques, only a regular pulse is generated in watermark extraction process to state the presence or absence of the watermark which limited the watermark's payload. However, some video watermark applications require high payload which is easily reachable due to the size of a video which is much higher than an image. Although some algorithms can embed a single message in all the video frames [3], the whole video bandwidth can be used by embedding multi-message in each video frame [4]. Using whole video bandwidth can degrade the imperceptibility and robustness due to embed watermark in less video samples. In conclusion, same or different embedding information can be used for each video frame which is totally application dependent.

5.3.1.1 Transform Domain

In transform domain, a transformation has applied to transform video frames into other domain. Then, spread spectrum (SS) is applied to embed the watermark into the transformed video stream. Two major techniques are JAWS and DWE.

JAWS: Just Another Watermarking System
JAWS video watermarking technique has been developed at the beginning of twentieth century in Philips research center [3]. The main application of this technique is for broadcast monitoring. This technique embeds the watermark in local activity of each frame. The watermark data are encoded based on shifting two reference patterns. To achieve this, a reference watermark W_r is constructed based on reference pattern P_r (which is produced based on a secret key) and binary message m as expressed in Eq. (5.1):

$$W_r = P_r - shift(m, P_r)$$ (5.1)

where shift(.) is a cyclically shifted function. Next, the watermarked frame $\hat{F}$ is constructed based on adding a watermark W into each original video frame F as in Eq. (5.2). In order to preserve imperceptibility, the watermark is adjusted based on two factors including local activity $\lambda(i)$ and a global embedding strength S.

$$\hat{F}(i) = F(i) + S \times \lambda(i) \times W(i)$$ (5.2)

In the extraction process, all possible reference patterns P_r are detected in frequency domain based on symmetrical phase only matched filtering (SPOMF). Finally, the watermark is extracted based on Eq. (5.1).

DEW: Differential Energy Watermark
DEW watermarking technique has been originally developed for image watermarking which is extended for video watermarking. This technique embeds the watermark into the MPEG video stream by watermarking I-frames [5]. In this technique, n blocks of pixels with size of 8×8 are selected to form

pseudorandomly secret key. Then, DEW technique has embedded the watermark bit into the high frequency of the DCT coefficients by differentiating the energy between bottom half and top half of each block. Therefore, this technique is known as deferential energy watermark.

5.3.1.2 Temporal Dimension Video Watermarking

Considering video stream as independent sequence of still image may have some shortcoming. For instance, temporal dimension is not taken into account by image-based video processing. However, temporal dimension such as perceptual shaping and motion predication which are incorporated by video coding community can be applied for video watermarking. Although many image watermarkings based on HVS can be used for video watermarking in frame-by-frame fashion, ignoring motion as a temporal dimension in video watermarking prevents to obtain optimum imperceptual video watermarking. On the other words, watermarking based on HVS properties for image such as contrast masking, luminance masking, and frequency masking cannot provide enough invisibility for video watermarking. Some extra sensitivities in human eye should be considered to develop video watermarking techniques [6]. This section investigates temporal video information by handling it into three parts which is discussed in the following:

Video Watermarking in a 1D Signal
Looking at video content as 1D signal without considering any dimensionality information such as spatial and temporal is one of the basic video watermarking techniques [7]. In this technique, 1D signal is formed by spatiotemporal scanning of the video content which is known as line-scanning strategy. Then, one of the 1D watermarking techniques like SS watermarking technique is applied to embed the watermark into the line-scanned video signal.

Video watermarking in a Temporal Signal
As discussed, ignoring temporal dimension can degrade the invisibility and robustness of the developed video watermarking technique. Thus, spatiotemporal dimensions must be considered in order to develop an optimum video watermarking technique. For this purpose, some key-dependent specific pixel locations are selected in pure temporal signal of the video content for watermarking [8]. These pixel locations should be selected based on video texture to ensure watermark invisibility. To achieve this aim, the pixels are selected where motionless border or rapidly changed pixel with time is available. However, the watermark capacity is not much due to less amount of these pixels. To solve this problem, temporal wavelet decomposition is applied on whole video content to provide more capacity for watermarking [9]. In temporal wavelet decomposition technique, video content is decomposed into motionless (static) and motion (dynamic) parts which allows multiple temporal scale watermarking. For example, DC frequency in wavelet frame has the lowest temporal video information. Independent component analysis (ICA) is another technique for generating frames with independent sources for watermarking.

Video Watermarking in 3D Signal

Various slices of a medical scan can be imaged into a sequence of video frames which provide homogenous three-dimensional video content. Many 3D transformation such as 3D DFT [10], 3D wavelet transform [11], and 3D Gabor transform [12] can be exploited for video watermarking by embedding the watermark into mid-frequency region to provide excellent robustness and imperceptibility. However, modeling the video based on one temporal dimension and two spatial dimensions as 3D signal modeling is not accurate enough for video watermarking. As a result, 3D signal techniques have very specific applications in video watermarking.

5.3.2 Video-Based Video Watermarking

Currently, video content is encoded based on conventional video compression standard like MPEG. The compressed form of the video content is distributed across the networks for consuming the bandwidth requirement. Thus, video content is usually compressed based on video elements including video bitstream, video structure, and motion compensation. Video-based watermarking approach applies these video elements to embed the watermark data in different compression levels. Since standard video compression techniques have applied transformation domain (e.g., MPEG which is based on DCT), some video-based watermarking techniques utilized certain legacy of the image-based watermarking techniques. In addition, embedding the watermark into code words before entropy coding can also be applied by some video watermarking techniques. As a result, embedding the watermark into code words (compressed format) and motion information is two available approaches for video-based watermarking. In the following, both techniques are discussed.

5.3.2.1 Video Watermarking Based on Compression Format

Integrating video watermarking based on video coding structure is a primary goal in compression format video watermarking technique. This integration attempts to reduce the computational complexity and provide optimum framework for video watermarking technique. In this technique, motion information and transform coefficients (e.g., DCT coefficients) in video encoder side are quantized by a vector or scalar quantization. Then, this quantized information is forming final video bitstream by using an entropy encoder. For instance, DCT coefficients in MPEG standard are scanned in zigzag order to form tuples (run, level) which are the number of zeros preceding a coefficient and the value of the quantized coefficient, respectively. Then, some lookup tables (as an entropy encoder) in the form of a variable length cod (VLC) are applied to convert tuples to code words. Consequently, embedding the watermark into these VLC code words can be performed to consume compression and decompression times. By embedding the watermark directly into video bitstream, imperceptibility is also preserved [5]. In order to encounter synchronization problem in VLCs, a new version named as reversible VLCS (RVLCs) is developed based on two-way decodable approach to provide resynchronization properties [13]. The ability of RVLCs for recovering the error makes this approach as core techniques for developing reversible video watermarking.

5.3.2.2 Video Watermarking Based on Motion Information

In order to reduce temporal redundancy in video content, motion information has coded by video compression standards. This motion vector is capable of video encoder/decoder to predict next frames based on current frames. In MPEG compression standard, three frames I, B, and P are applied. Embedding the watermark data in motion vectors is one of the interesting ways for video watermarking [14]. For example, motion vector may be quantized to an odd value when the watermark bit is equal to 1 and it must be quantized to an even value when the watermark bit is equal to 0. To achieve high watermark imperceptibility, the motion vectors with high magnitude can be selected for quantization to reduce the watermark's distortion. Furthermore, there is a possibility to embed the watermark into the angle between vertical and horizontal components of the motion vector [15]. In this line, some motion vector quantization techniques including an angular grid, circular grid, and square grid have been proposed [16]. Although video watermarking based on motion information is highly robust, predicting the perceptual degradation impact is a major challenge. For instance, partial encryption technique called waterscrambling can be applied for motion vector quantization to degrade watermark's invisibility effect to trigger an impulsive buying action [17].

5.4 Video Watermarking Attacks

Similar to image watermarking, video watermarking is targeted to multiple attacks which can take place with various purposes. Although some of the attacks between image watermarking and video watermarking are common, it is critical for video watermarking developers to know specific video to share their knowledge to break the system [18]. This attack can be extended for video watermarking for obtaining unprotected video content. In the collusive attack, the number of colluder users c is always less than the whole number of user u ($c < u$). In fingerprinting application of the watermark, video content is distributed among some audience who are concerned about their copyright. Thus, different copies are assigned to each customer instead of assigning same item to all of them. As a result, each video content can be traced by random and imperceptible tracers which are embedded into the video content. However, collusive attacks have been successfully applied for fingerprinting in digital watermarking [19].

5.4.1 Temporal Frame Averaging (TFA)

TFA attack is applied when uncorrelated watermark is embedded into the video frames. According to a fact that neighborhood video frames are most probably similar, a temporal low-pass filter is applied for averaging video frames to construct unwatermarked video content. Although TFA attack is proper for static scene, the poor quality result for dynamic scene can be improved by frame registration.

5.4.2 Watermark Estimation Remodulation (WER)

WER attack is mainly developed for dynamic scenes which estimate the watermark blindly without accessing to original video content. For this purpose, the hidden watermark is estimated based on various video frames. Then, this estimation is refined during receiving more video frames. Lastly, the watermark is removed from the content.

5.4.3 Watermark Estimations Clustering and Remodulation (WECR)

WECR attack is an enhanced version of WER attack which is using some key frames in watermark video content instead of using whole sequence of the video frames. To achieve prefect watermark removal, k-means algorithm is applied to construct N centroids which reveal the watermark patterns.

5.4.4 Watermarking Subspace Estimation Draining (WSED)

WSED attack is an enhanced version of WECR which applies subspace to project the extracted watermark (estimates). For this purpose, each centroid C_i is retrieved by identifying the coordinates of each cluster in watermarking subspace.

5.4.5 Block Replacement (BR)

BR attack has used a special kind of temporal redundancy which is known as self-similarities inside each video frame to replace the input signal with similar desired signal. Although using spatial self-similarities for replacing the input signal is efficient, the replaced block cannot vary with watermarked input signal to preserve the imperceptibility of the watermarked video. Therefore, BR attack should be wised enough to provide trade-off between imperceptibility and attack efficiency by using intensive testing. In addition, the replaced block should be far enough to input signal because the watermark may be available in watermarked block.

5.4.6 Block Combination

Due to imperceptibility problem in replacement block attack, another attack known as block combination is designed to guaranty imperceptibility and quality of the watermarked video. Two approaches have existed for selecting the number of blocks including fixed approach and adaptive approach. In the fixed number of the block attack, the optimal replacement block is computed based on a threshold; somehow the quality of the video is preserved. This is performed by choosing the most similar blocks. In the adaptive number of the block attack, some steps in fixed approach are improved with checking the amount of distortion which is injected by candidate block for replacement and is bounded between a minimum and a maximum threshold.

5.4.7 Block Projection

Block combination attack cannot guarantee the optimum block replacement due to restriction of the block subset to a minimum threshold and a maximum threshold. Although block combination can compute suboptimum blocks, it cannot compute all possible combination of the blocks due to computational complexity. Block projections try to solve this limitation by generated subspace for target blocks based on orthogonal projections such as Gram–Schmidt orthonormalization and principal component analysis.

5.4.8 Block Restoration

In digital transmission, some blocks may be corrupted or lost which are recovered by error concealment technique through the neighborhood blocks of missed block. Block restoration attack is applying error concealment technique to enhance block replacement attack efficiently. Similar to block replacement attack, some blocks of the watermarked video are considered as missing information which must be concealed through error concealment technique [20]. However, two major limitations are as follows: lack of attack strength to adapt the block restoration attack based on watermarked strength and ignoring some blocks (which are not missed in reality) caused to degrade the video quality.

5.4.9 Block Swapping

In contrast to random bending attack [21] that is not modified geography of the watermarked information, block swapping attack tries to shuffle the watermark sample by preserving synchronization among the watermarked data. On the other hand, each video block is replaced with similar block in order to alter the geography of watermarked video. Consequently, watermark extraction process is impossible because resynchronization is not longer possible.

5.5 Video-Enforced Strategies

As discussed, many techniques are available to embed the watermark into the video content. However, embedding video sequence as still images is the most common strategy [22]. Furthermore, inserting the watermark based on spread spectrum (SS) theory is the simplest which can embed different or same watermark into the video frames. In the following, four well-known approaches are described.

5.5.1 SS System

SS system is the basic system for video watermarking which embeds a pseudorandom watermark into the video content (which is considered as 1D signal) frame-by-frame strategy. Although watermark imperceptibility may be improved by perceptual shaping, a global watermark strength in SS system may not able to preserve enough imperceptibility from flicker artifact.

5.5.2 SS-1 System

SS-1 system is developed in order to solve limitations of SS system. Actually, temporal desynchronization such as frame insertion and frame dropping can confuse the watermark detector in SS system. However, SS-1 system embeds same pseudorandom watermark in every frame of the video [3]. Therefore, SS-1 system requires less computation for extracting the watermark due to the use of linearity of the operator. On the other words, correlations between the same watermark and different video frames are computed for watermark detection which means a single correlation between the watermark and the average of all video frames is computed.

5.5.3 SS-N System

Although SS-1 system can overcome some temporal synchronization attacks, embedding same watermark has some vulnerabilities which may give an adverse opportunity to group frames with same watermark and then apply a WER attack to estimate the watermark. SS-N system is secured SS-1 system by selecting a watermark randomly for each video frame. The watermark selection is based on a finite set (N) of reference watermarks W_i. The new architecture for SS-N is enabling to state both SS system and SS-1 system as specific cases of SS-N system where $N = \infty$ and $N = 1$, respectively. Similar to previous systems, the absolute values of N linear correlations should be added before being temporally averaged to extract the watermark. Due to the use of absolute values, linearity of the operator cannot be used. As a result, N can define the complexity of the watermark extraction process. Moreover, synchronization is not needed for extracting watermark in SS-N system.

5.5.4 SS-α System

SS-α system adopted the SS-N system by providing embedding strength which may be a time-dependent function like a sinusoid modulation function. Therefore, N reference watermark patterns are linearly combined somehow coefficients are mixed temporally by embedding strength function. Three main constraints must be satisfied for embedding strength function. First, the embedding strength function must immune to TFA attacks by varying smoothly with time. Second, the embedding strength function must resist against WER attacks by keeping zero mean. Third, the embedding strength function must resist against WECR attacks by discrete sampling of the video frames to a large number of values. Unlike some watermarking systems that use embedding strength for improving imperceptibility, the main aim of embedding strength for SS-α system is security issues. Whenever the amount of α is increased, the amount of watermark distortion is increased.

5.5.5 Discussion

SS system is distributing the embedded watermark on the unit sphere uniformly. However, the watermark is concentrated in single narrow area on the unit sphere in SS-1 system. Even for SS-N system, the watermark is distributed among some clusters. For example, for $N = 4$, four clusters are gathered on unit sphere. Although both SS system and SS-1 system are easily degraded by collusive attacks, fortunately a trajectory on the unit sphere, which is defined by successive watermark, can be defined to improve robustness against collusive attacks. For WECR attacks, accumulation points should be contained in trajectory. Also, trajectory must be continuous by averaging successive watermarks resulting in a watermark near the surface of the unit sphere.

5.6 Critical Review in Video Watermarking

Table 5.1 summarizes the evaluation of the performance of each watermarking technique in terms of robustness, imperceptibility (invisibility), and capacity. As seen, none of them is ideal due to the nature of watermarking which is a trade-off among these criteria. Whenever a criterion is increased, other criteria are decreased. Although the result for the proposed watermarking systems seems to be good, they are still reported under their assumption. In real environment condition, there is still a huge gap to use an efficient watermarking technique to embed the voice feature in digital image.

Table 5.1 Comparison of related image and video watermarking methods

WM method	Capacity (No of WM bits)	Robustness (BER %)	Imperceptibility (PSNR)	Blindness
Image watermarking methods				
Histogram [23]	25	0.003	48.34	Semi
Spatial domain [24]	256	0.002	47.54	Non-blind
DWT-DCT-SVD [25]	256	0.001	113.42	Semi
DWT-DFT phase [26]	256	0.050	76.11	Blind
DFT [27]	256	0.260	60.52	Blind
SVD [28]	256	0.001	73.65	Blind
DWT [29]	64	0	44.25	Blind
DWT [30]	512	0	48.07	Blind
RDWT [31]	4096	0.35	41.53	Blind
Alternative video watermarking methods				
DCT [32]	256	0.004	60.24	Semi
Wavelet packet [33]	6336	0.001	64.87	Blind
Ridgelet [34]	2304	0.03	44	Blind
DWT+ NN [35]	2304	0.02	39.08	Blind
DWT+ spatial [36]	2304	0.05	N/A	Blind
DWT [36]	2304	0.11	40.07	Blind
DWT [37]	2304	0.15	38.95	Semi
DWT+ QIM [37]	1024	0.02	45.25	Blind

Table 5.2 Advantage and disadvantage of each video watermarking technique

Video Watermarking technique		Advantage	Disadvantage
Image-based	Transform domain	Can be Inherited from image watermarking	Complex computationally
	Temporal dimension	Highly robust	Complex computationally
Video-based	Compression format	Real-time	Depended on video format
	Motion information	Highly robust	Less fidelity and imperceptibility

Image and video watermark techniques can be similar in some ways, but they are different from other ways. Thus, image-based and video-based watermarking techniques have their properties. The pros and cons of each video watermarking technique are summarized in Table 5.2.

References

1. Bodo, Y., N. Laurent, and J.-L. Dugelay. 2003. Watermarking video, hierarchical embedding in motion vectors. In *Proceedings of the 2003 international conference on image processing, ICIP 2003*. IEEE.
2. Barni, M., et al. 2000. A robust watermarking approach for raw video. In *Proceedings of the 10th international packet video workshop*.
3. Kalker, T., et al. 1999. Video watermarking system for broadcast monitoring. In *Electronic imaging'99*. International Society for Optics and Photonics.
4. Dittmann, J., M. Stabenau, and R. Steinmetz. 1998. Robust MPEG video watermarking technologies. In *Proceedings of the sixth ACM international conference on Multimedia*. ACM.
5. Langelaar, G.C., R.L. Lagendijk, and J. Biemond. 1998. Real-time labeling of MPEG-2 compressed video. *Journal of Visual Communication and Image Representation* 9(4): 256–270.
6. Kim, S.-W., et al. 1999. *Perceptually tuned robust watermarking scheme for digital video using motion entropy masking*.
7. Hartung, F., and B. Girod. 1998. Watermarking of uncompressed and compressed video. *Signal Processing* 66(3): 283–301.
8. Martin, X.N., M. Schmucker, and C. Busch. 2002. Video watermarking resisting to rotation, scaling, and translation. In *Proceedings of the SPIE security watermarking of multimedia contents IV*. Citeseer.
9. Swanson, M.D., B. Zhu, and A.H. Tewfik. 1998. Multiresolution scene-based video watermarking using perceptual models. *IEEE Journal on Selected Areas in Communications* 16(4): 540–550.
10. Deguillaume, F., et al. 1999. Robust 3D DFT video watermarking. In *Electronic imaging'99*. International Society for Optics and Photonics.
11. Kucukgoz, M., et al. 2005. Robust video watermarking via optimization algorithm for quantization of pseudo-random semi-global statistics. In *Electronic imaging 2005*. International Society for Optics and Photonics.
12. Zhang, L.-H., H.-T. Wu, and C.-L. Hu. 2004. A video watermarking algorithm based on 3D Gabor transform. *Journal of Software* 15(8): 1252–1258.
13. Mobasseri, B.G., and D. Cinalli. 2004. Reversible watermarking using two-way decodable codes. In *Electronic imaging 2004*. International Society for Optics and Photonics.

14. Jordan, F. 1997. *Proposal of a watermarking technique for hiding/retrieving data in compressed and decompressed video.* ISO/IEC Doc. JTC1/SC 29/QWG 11 MPEG 97/M 2281.
15. Zhang, J., J. Li, and L. Zhang. 2001. Video watermark technique in motion vector. In *Proceedings of XIV Brazilian symposium on computer graphics and image processing.* IEEE.
16. Yann, B., L. Nathalie, and D. Jean-Luc. 2004. A comparative study of different modes of perturbation for video watermarking based on motion vectors. In *Proceedings of the 12th European signal processing conference.* IEEE.
17. Bodo, Y., et al. 2004. Video waterscrambling: Towards a video protection scheme based on the disturbance of motion vectors. *EURASIP Journal on Advances in Signal Processing* 2004(14): 1–14.
18. Menezes, A.J., P.C. Van Oorschot, and S.A. Vanstone. 1996. *Handbook of applied cryptography.* Boca Raton: CRC press.
19. Wu, M., et al. 2004. Collusion-resistant fingerprinting for multimedia. *IEEE Signal Processing Magazine* 21(2): 15–27.
20. Cayre, F., et al. 2003. Application of spectral decomposition to compression and watermarking of 3D triangle mesh geometry. *Signal Processing: Image Communication* 18(4): 309–319.
21. Petitcolas, F.A., R.J. Anderson, and M.G. Kuhn. 1998. Attacks on copyright marking systems. In *Information hiding.* Berlin: Springer.
22. Doerr, G., and J.-L. Dugelay. 2003. A guide tour of video watermarking. *Signal Processing: Image Communication* 18(4): 263–282.
23. Zong, T., et al. 2014. *Robust histogram shape based method for image watermarking.*
24. Cheung, W. 2000. Digital image watermarking in spatial and transform domains. In *Proceedings of the TENCON 2000.* IEEE.
25. Navas, K., et al. 2008. DWT-DCT-SVD based watermarking. In *3rd international conference on communication systems software and middleware and workshops, COMSWARE 2008.* IEEE.
26. Kang, X., et al. 2003. A DWT-DFT composite watermarking scheme robust to both affine transform and JPEG compression. *IEEE Transactions on Circuits and Systems for Video Technology* 13(8): 776–786.
27. Li, J., et al. 2009. An adaptive secure watermarking scheme for images in spatial domain using fresnel transform. In *Information science and engineering (ICISE).* IEEE.
28. Ghazy, R.A., et al. 2015. Block-based SVD image watermarking in spatial and transform domains. *International Journal of Electronics* 102(7): 1091–1113.
29. Lin, W.-H., et al. 2008. An efficient watermarking method based on significant difference of wavelet coefficient quantization. *IEEE Transactions on Multimedia* 10(5): 746–757.
30. Ma, B., et al. 2014. Secure multimodal biometric authentication with wavelet quantization based fingerprint watermarking. *Multimedia Tools and Applications* 72(1): 637–666.
31. Agarwal, H., B. Raman, and I. Venkat. 2014. Blind reliable invisible watermarking method in wavelet domain for face image watermark. *Multimedia Tools and Applications*: 1–39.
32. Wang, Y., and A. Pearmain. 2006. Blind MPEG-2 video watermarking in DCT domain robust against scaling. *IEE Proceedings-Vision, Image and Signal Processing* 153(5): 581–588.
33. Bhatnagar, G., and B. Raman. 2012. Wavelet packet transform-based robust video watermarking technique. *Sadhana* 37(3): 371–388.
34. Khalilian, H., S. Ghaemmaghami, and M. Omidyeganeh. 2009. Digital video watermarking in 3-D ridgelet domain. In *11th international conference on advanced communication technology, ICACT 2009.* IEEE.
35. Li, X., and R. Wang. 2007. A video watermarking scheme based on 3D-DWT and neural network. In *Ninth IEEE international symposium on multimedia workshops, ISMW'07.* IEEE.
36. Huai-Yu, Z., L. Ying, and W. Cheng-Ke. 2004. A blind spatial-temporal algorithm based on 3D wavelet for video watermarking. In *2004 IEEE international conference on multimedia and expo, ICME'04.* IEEE.
37. Yassin, N.I., N.M. Salem, and M.I. El Adawy. 2014. QIM blind video watermarking scheme based on wavelet transform and principal component analysis. *Alexandria Engineering Journal* 53(4): 833–842.

Chapter 6
Three-Dimensional (3D) Watermarking

6.1 Introduction

Three-dimensional (3D) computer graphic is applied widely in digital archives, video games, entertainment, animation, MPEG4, and Web3D. 3D watermarking focuses on embedding hidden data in 3D materials. As discussed in Chap. 5, 3D watermarking is very similar to a compression standard that implies watermarks on a sequence of still images [1]. 3D watermarking can be seen from another aspect that embeds a watermark in a three-dimensional transform. The definition and characteristics of 3D watermarking as a 3D transform and as a data-hiding technique for digital video are basically different; thus, they are discussed in two separate chapters. In this chapter, the latter definition is favorable.

The philosophy of 3D watermarking is same as the other types of watermarking. The goal is to embed a confidential data in a 3D graphical model and transfer the watermarked data to the destination on the condition that the watermarked and the original 3D objects could not be distinguished visually by the user. 3D watermarking was initially introduced in [2] and still is an ongoing and fresh area of research. So far, many watermarking schemes have been proposed for image and video watermarking. However, 3D graphic differs from the 2D and 1D graphical objects, and traditional watermarking techniques are not applicable in 3D graphical models. The sampling of a 3D signal is not possible with conventional sampling devices and algorithms as used for 2D or 1D signals. Moreover, unlike an image or an audio signal, the result of 3D sampling could not be illustrated in the form of a single dataset. Spectral analysis functions that are essential for conventional sampling and analysis of signals including Fourier transform (FT), discrete cosine transform (DCT), and wavelet transform (WT) become very sophisticated in 3D sampling.

There are many different ways for the presentation of a 3D signal on both geometry and connectivity; e.g., a mesh can be denoted as a 1-to-4 connectivity (one vertex is connected with four neighbors), or 1-to-6 connectivity. Both geometry and

© Springer Science+Business Media Singapore 2017
M.A. Nematollahi et al., *Digital Watermarking*, Springer Topics
in Signal Processing 11, DOI 10.1007/978-981-10-2095-7_6

connectivity are not reliable and powerful for efficient sampling of a 3D graphical object. In some cases [3, 4], the regular sampling is assumed for regular connectivity, and in other sampling techniques [5], a small number of vertices conserves the shape of the mesh surface. Furthermore, if a uniform sampling utilizes for sampling, the result would be totally new samples.

The components of conventional signals are ordered simply; e.g., an image can be ordered in terms of the elements located in the columns or rows, or elements of an audio signal are ordered based on the time of occurrence. Nevertheless, there is no specific rule or way for a stable order sequence of 3D signals. Therefore, an explicit way for sequence order of a 3D object is not available, either for connectivity or for geometry. Thus, the development of available 3D spectral watermarking schemes (e.g., [6]) is hard-hitting.

In this chapter, firstly, different concepts and definitions are discussed in 3D model. Secondly, background of 3D watermarking is explained. Thirdly, various techniques for 3D watermark are described. Fourthly, available attacks which degrade the 3D watermarking are discussed. Lastly, distortion evaluation techniques for validating the performance of 3D watermarking are explained.

6.2 3D Modeling Representation

3D objects are modeled by the collection of points. The anticipated shape of the modeler is commonly represented in form of a polygon in 3D models [7]. A set of surfaces construct an object, and a group of polygons make surfaces as shown in Fig. 6.1. The shape of a polygon is formed by the intersection of straight lines called as edges. The intersection points are named as vertices, and the area limited inside the polygon is the face [8]. Obviously, vertices are 3D points located in the world coordinate space. There are two faces for a polygon: front and back. The polygon borders separate the inside region from outside one, and the objects that form polygonal model are assembled by primitives. Collection of the adjacent polygons builds polygonal meshes. As 3D models are described by geometric and topological boundaries, polygon representation was formerly called as boundary representation.

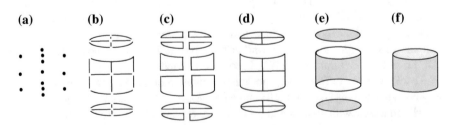

Fig. 6.1 Various component in a mesh model. **a** Vertices. **b** Edge. **c** Faces. **d** Polygons. **e** Surface. **f** 3D object

For processing purposes in applications, e.g., 3D model-based video coding [9], digital terrain modeling [10] and scientific visualization [11] polygonal surfaces are needed to be estimated. Despite the expectations, obtaining a polygonal representation of 3D graphical objects is not without difficulties. Moreover, the acceptable accuracy of modeling (i.e., the difference between the curved surface of an object and the faceted representation) is tentative [12].

Polygonal representation is not the sole approach for modeling 3D signals. There are some other approaches including non-uniform rational basis spline (NURBS) and primitive modeling. A widespread model for curves and surfaces in computer graphics is NURBS [13]. A curve is constructed by set of points, and the line that describes the curve is counted as spline. In this model, a mesh is made by intersection of splines. The intersecting splines make areas called as patches. Another approach for modeling complex objects is primitive modeling. This approach combines some primitive objects and shapes, e.g., sphere, cube balls, cylinders, and cones (or cubes), and modifies their shapes to achieve favorite objects.

For converting complex objects to meshes, the help from 3D software is necessary. The primitives can be utilized directly to be rendered by software or can be kept for further operations and rendering. Modeling process can be done by a particular program (e.g., from Z, Maya, 3DS Max, Blender, Lightwave, Modo, and solid Thinking) or an application component (Shaper, Lofter in 3DS Max) or some scene description language (as in POV-Ray). Nevertheless, in some cases, distinguishing between different phases is not simple, and other processes need to be joined (this is the case, e.g., with Caligari trueSpace and Realsoft 3D).

In real systems, objects are sometimes more complicated, e.g., blowing sand, clouds, and liquid sprays. For modeling these objects, a mass of 3D coordinates is necessary where multiple polygons, splines, texture splats, and points are operated.

6.3 Background of 3D Watermarking

It is understood that 3D objects can be modeled by using polygonal meshes, constructed solid geometry, voxels, or implicit group of parameterized equation as in NURBS and simply and with high quality are duplicated, transferred, and stored. Almost all models can be modified to triangular meshes; therefore, it is the dominant approach for 3D modeling in computer graphics. Henceforth, the triangular mesh is considered for the representation of 3D models. Given a triangular mesh named as object O, where a set of vertices, i.e., $V = \{vi \ 2 \ O | i = 1, \ldots, |O|\}$, describes its geometry, |O| and |F| denote the number of vertices and the number of faces of O, respectively. F is the set of triangles that link the vertices and is defined as $F = \{Fi \ 2 \ O | i = 1, \ldots, |F|\}$. Although vertices or faces can have different colors, textures, and shading, they are simply intruded by attackers and thus are not appropriate candidates for embedding the watermark. A non-boundary edge of object O must solely join two neighboring faces. Table 6.1 delineates the common 3D mesh models and its characteristics from LIRIS database [14].

Table 6.1 The most common 3D mesh models

Model name	Diagonal lengths of the bounding boxes	Number of vertexes	Figure
Bunny	2.131131	34,835	
Dragon	2.364309	50,000	
Venus	2.753887	100,759	
Horse	2.104646	112,642	
Rabbit	1.823454	70,658	
Ramesses	1.853616	826,266	
Cow	2.032768	2904	
Hand	2.157989	36,619	

(continued)

Table 6.1 (continued)

Model name	Diagonal lengths of the bounding boxes	Number of vertexes	Figure
Casting	2.322810	5096	
Crank	2.423323	50,012	

6.4 Attacks in 3D Watermarking

A 3D watermarking is subject to different types of attacks. An attacker may attempt to obtain confidential data, modify the data, or replace his/her data with the watermark to mislead the receiver or misuse the confidential information. Therefore, it is essential for 3D watermarking method to be resilient to attacks. For this purpose, the knowledge on different types of attacks and the consequences is required. 3D attacks are generally classified into two categories [15]. Some attacks do not change the connectivity or the geometry of the mesh model and thus are not destructive for the 3D model and are simply tolerated by watermarking techniques, e.g., rotation, uniform scaling, mesh registration, geometry transformation, and object description file shuffling. This category is known as distortionless attacks. Indeed, another category of attacks refers to distorting attacks that destruct the connectivity or the geometry or both. Major attacks in 3D watermarking are summarized in Fig. 6.2.

As the 3D model gets complicated, the impact of the attack becomes more complex and unpredictable. A mesh model consists of irregular shapes and unorganized points, with variable topology and connectivity methods; it is not easy to imagine the spots that an attack may alter. Definitely, when there are several spaces for an attack to be imposed, the prevention from the attack becomes sophisticated. Major distorting attacks are counted as follows.

- Translation/rotation/uniform scaling: Geometric transformation is widely applied to embed a 3D graphic model in a scene especially in computer graphics.
- Noise. A noise attack harms the vertices of a mesh in random fashion.
- Retriangulation. It alters the connectivity of the vertices in a mesh model.
- Mesh smoothing. A filter is applied on the mesh model and makes the surface of a polygon smooth, e.g., Taubin filter [16] that reduces the jagged surface by performing a low-pass filter.
- Polygonal simplification. It simplifies the mesh model by eliminating the non-salient faces in the form of a low-level version.

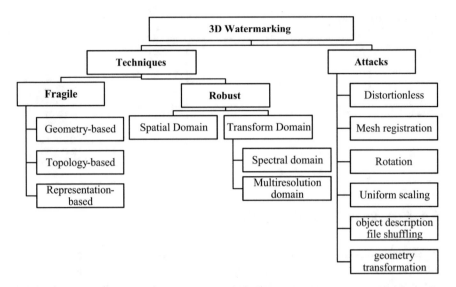

Fig. 6.2   Summary of 3D watermarking and attacks

- Cropping. The undesired parts (from the point of an attacker) of the model are removed by the attacker. That can be a hand of a statue, for example.
- Remeshing. A geometric resampling approach changes the mesh model by redefinition of the vertices' connectivity of the 3D mesh model.

6.5   Techniques in 3D Watermarking

In contrast to audio, image, and video watermarking, a few researches have been conducted for 3D watermarking. However, these researches can be classified into two main groups including fragile 3D watermarking techniques and robust 3D watermarking techniques [17]. Although fragile 3D watermarking tries to improve the payload which is a requirement for one-to-one application, robust 3D watermarking attempts to improve the robustness against one-to-many application. In the following, both 3D watermarking techniques are fully described.

6.5.1   Fragile 3D Watermarking

According to Fig. 6.2, there is a category of 3D watermarking systems that implies fragile watermarking. Fragile watermarking techniques consist of three main techniques including geometry-based, topology-based, and representation-based. The topology-based and geometry-based techniques modify the features of a 3D mesh model, while redundancy is added to the indexed representation of the model for the representation-based fragile watermarking. In the following, each of these techniques is discussed separately.

6.5.1.1 Geometry-Based Watermarking

The most common approach for 3D watermarking is geometry-based that inserts hidden data into the geometry of the mesh model. Two methods for watermarking technique in this category, i.e., triangle similarity quadruple embedding and tetra-hedral volume ratio embedding, were first introduced in [18]. Then, a blind technique for 3D watermarking was presented in [19]. According to the mesh model, it firstly generates a list of triangles as the candidates for carrying the hidden data and then alters each candidate for the bit of hidden data that should be carried-on. As one bit is embedded per vertex, the capacity of hidden data is low. In this method, each triangle is labeled with two states, either 0 or 1, which depends on the position of the triangle summit of the vertex regarding to the projection of the opposite edge. This method is shown in Fig. 6.3. For a better hiding of data, the projection of a vertex is moved toward the nearest correct interval of its opposite edge. Initially, a triangle with the greatest area or the lowest area is selected from the list, and the remaining triangles are ordered in the same way. Otherwise, a

Fig. 6.3 A vertex is projected to nearest interval for binary watermark bits (**a**) an edge is divided into two intervals for better robustness (**b**) an edge is divided into four intervals for better imperceptibility

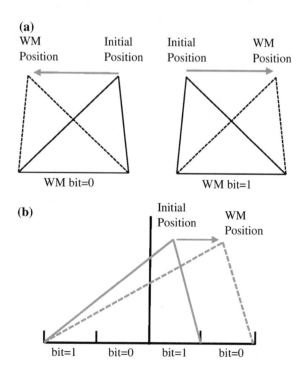

triangle among the six available triangles of the mesh model, intersected with the three main axes located at the center of the gravity index, is chosen.

PCA is applied in geometry-based and topology-based techniques to degrade the synchronization problem as in [20]. It further improves the robustness of 3D watermarking by tolerating the content-preserving operations except 3D simplification and remeshing attacks and other malicious attacks. As the opposite of a summit of a vertex is required to be recognized, it is not applicable in the approaches with no triangular mesh structure and random topology.

A multi-level embedding process was introduced in [21] that improved the low capacity of hidden data in 3D triangular mesh [19] to 3 bits per vertex. The features and limitations of human visual system (HVS) were applied for higher message insertion in the mesh model. Moreover, the execution time of this algorithm is low for both procedures of embedding data and data extraction that makes it appropriate for large 3D datasets. HVS recognizes the changes on the smooth surfaces simpler than jagged or rough surfaces. This feature was applied in [22]. The intensive level of rough surfaces has different impacts on the visibility of HVS. The invisibility of hidden data in high fluctuated surfaces is high; therefore, a higher rate of hidden data is possible to be embedded in these areas. The reverse concept can be imagined for lower intensity of roughness. The algorithm in [22] changes the level of data hiding according to the intensity level of a surface's roughness; therefore, it is called a flexible watermarking approach. It extracts the correlation between neighboring polygons to estimate the level of roughness or smoothness. Flexible 3D watermarking upgrades the capacity of hidden data to more than that in polygen model [21]. It reduces the distortion of hidden data, as well. Due to machine precision errors, the performance of this method (in terms of capacity) is low for the models with few vertices.

PCA was applied to construct a new coordinate system comprising the three main axes [23]. This idea was replaced with the former two-state triangle models. As shown in Fig. 6.4, two extreme end points, V_a and V_b, are the furthest projections of the mesh vertices on the new x, y, or z axis and determine the line segment. Then, the line is uniformly divided into a set of intervals (segments), while each interval denotes a two-state object in an interleaved manner.

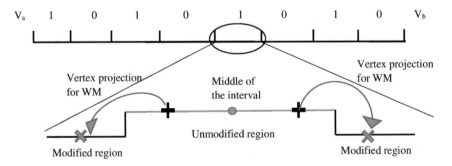

Fig. 6.4 A PCA classification approach for 3D watermarking

A segment is generally divided into two main regions including changed region and unchanged region. The message data is embedded in the portions of the segment depending on the inserted bit value and the state of the interval into which the vertex is projected. A to-be-embedded vertex can be kept unmodified or can be moved into the change region of the interval for embedding data. An approach to minimize the distortion on the model is to move the vertex to the nearest projected position on the interval; i.e., when the projection is located on the right-hand side of the interval center, the vertex is moved to the right into the change region, and vice versa. This method is shown in Fig. 6.4.

It is concluded that geometry-based watermarking improves the capacity of hidden data with a low distortion on the model. Nevertheless, it can tolerate transform attacks but shows a low resistance against transform attacks. Furthermore, it does not tolerate malicious attacks.

6.5.1.2 Topology-Based Watermarking

Topology-based watermarking methods modify the connectivity or topology of a mesh model to hide the data. A limited information can be carried by the connectivity, and there are few topology-based watermarking techniques in the literature. One of the connectivity-based methods applies a triangle subdivision for data embedding [24]. For each three edges of a triangle, a new vertex is added. Once any two vertices are connected, the triangle is divided into four subtriangles. It is equal to the condition where a longer triangle edge is divided into two smaller edges. A length-based ratio is calculated by considering the length of two new introduced edges, and then, one bit is embedded based on the obtained ratio. This approach is considered as blind because there is no need for original signal during extraction of the hidden data.

A lossless 3D watermarking, based on the topology, was presented that modifies the connectivity by using a minimum spanning tree (MST) [25]. It firstly generates a MST to cover the vertices of the 3D model. The MST is scanned to detect the candidate mesh area for embedding the hidden data. When the mesh is reorganized, the PCA ensures that the location of starting vertex is not modified and it depends on the secret key solely. At the last step, by means of modifying the connectivity, the embedded bit message is carried out in the selected area of the mesh. As the coordinates or the location of vertices in the mesh model are fixed, this method is considered as distortion free.

6.5.1.3 Representation-Based Watermarking

Redundancy is the main concept in representation-based 3D watermarking. As the original geometry and connectivity are not manipulated for data hiding, deployment of the redundancy in the representation-based watermarking methods makes them distortion free. The scope of representation-based watermarking is limited in comparison with other methods that embed information inside the mesh model, for that the number of methods in this category is low.

Regarding the traversal orders, a method in representation-based category [26] changes the order of the vertex representation and the polygon representation for embedding hidden data in a random fashion. As the order of the vertex and the polygon is altered solely in the list of vertices and the list of the polygons, the geometry of the mesh is kept intact. Therefore, this method is lossless.

A representation-based method permutes the order of storing faces and vertices in the indexed representation of the 3D mesh model for hiding the data [27]. The encoder and the decoder derive a reference ordering based on the connectivity of the 3D mesh model consistently. This reference order acts as the basis of the permutation with no modifications on the geometry of the 3D model, but degrades the rendering performance of the method. This limitation was resolved by increasing the capacity of hidden data, up to 0.63 bits per primitive, in average [28]. Another limitation was resolved by improving the mapping process [27]. Previous method encoded both the interior nodes and the leaf nodes. Thus, it presented the corresponding bit streams twice. On the other hand, the leaf nodes are solely encoded for mapping in [28].

6.5.2   Robust 3D Watermarking

Regarding the development of the domain of 3D watermarking, robust watermarking techniques are classified in two categories including spatial domain and transform domain. In the following, each domain is explained in details.

6.5.2.1   Spatial-Based Watermarking

Fragile watermarking is a special scheme of 3D watermarking that was introduced for authentication and attack localization. High sensitivity against minor and major modifications and localization and identification of the endured attacks are strongly expected by fragile watermarking. It should tolerate the content-preserving operations as well. However, robustness of fragile watermarking is not essential against malicious attacks.

Causality and convergence are two major difficulties during the data hiding in a 3D signal [29]. Alteration of neighborhood among the vertices that formerly have been processed may cause changes on the further processing of the neighboring vertices. It is called as causality and yields to unwanted modifications of extracted bits even in attack-free situations. When heavy distortion of the original model occurs and some vertices are not reached to a predefined relationship, convergence is recognized.

In a method in [30], 3D meshes were altered by fragile watermarking scheme to verify that two predefined hash functions generate same values when watermarking is applied to each vertex. This method suffers from the causality and convergence problems. It highly depends on the order of the vertices' traversal, does not distinguish the incidental data processing from the malicious attacks, and cannot localize the modifications. As a solution for causality problem in [30], a fragile watermarking scheme was proposed in [31] that resolves the causality problem. Regardless of the neighbors of a vertex, two different hash functions are performed on the vertex coordinates. As the order of vertices is not taken into account, the order-dependent attacks, e.g., vertex reordering, are simply tolerated. It is resilient to certain shape-preserving data processing. An authenticating 3D watermarking approach is performed on meshes in [32]. It finds the centroids of the mesh faces and adjusts their positions for embedding the watermark bits. Although this method is sensitive to some operations, it tolerates the uniform scaling, rotation, and translation.

A combination of the idea in [32] and [31] was proposed in [29] in order to solve the causality and convergence problems. For this purpose, different functions are applied during watermarking and the three coordinate components of a certain vertex are set. A marked vertex is indicated by a modulation on x coordinate, and watermarked information and the hash value of watermark are carried by y and z coordinates, respectively. This method keeps the barycenters of the marked vertices to intact and controls the average of distortion level during watermarking.

Integral invariants were utilized in a semi-fragile watermarking algorithm which tolerated certain noise attacks and severe transforms [33]. Each vertex has an integral invariant. When the position of a vertex and its neighbors are shifted, some of the vertices and their integral invariants are changed. This modification is utilized for data hiding through watermarking. Other fragile watermarking schemes are discussed in [34, 35].

The concern on the robustness of watermarking techniques is dominant in the literature. In line with this, a set of mesh vertices are arranged for embedding the watermark [36]. Only one bit is hided into each set. Adjustment of the groups is performed based on a secure key with an additive method that takes each group of vertices to the center of the model by modifying the distances. This method is the first global watermarking method that uses the geometric features. The strength of the watermarking scheme depends on the features of the mesh that are extracted locally and enhances the robustness but degrades the visibility of the watermarking. As the extraction procedure requires the original model, it is counted as a blind watermarking.

As the Cartesian coordinates are susceptible to attacks, some watermarking techniques utilize the spherical coordinates (ρ, θ, ϕ) rather than the Cartesian coordinates (x, y, z). The Euclidean distance between the mesh vertex and the barycenter is the radial component and is shown by ρ. ρ is a good candidate for carrying the watermark bits, as it does not alter under some operations, e.g., rotation and translation of 3D image, and almost, the values of ρ make the shape of a mesh which yields to its robustness.

There are two 3D watermarking schemes in spherical coordinate system that utilize the radial component ρ for embedding the watermark bits [37]. One method uses the mean, and the other utilizes the variance for embedding the watermark.

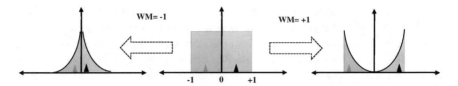

Fig. 6.5 3D watermarking based on histogram manipulation

Consider ρ_i as the Euclidean distance between vertex i and barycenter of the 3D mesh model and N as the total number of vertices of mesh model. Initially, a histogram is generated by $\rho_i | 1 \leq i \leq N$. The values of ρ_i are then normalized between 0 and 1, i.e., $0 \leq \rho_i' \leq 1$. The mean of the normalized values is ½, and the variance is near 1/3. Embedding a watermark bit is performed in a manner that the value of mean remains almost ½ and the variance almost 1/3. Both above-mentioned approaches are brought in Fig. 6.5.

As shown in Fig. 6.5, the histogram's mean value is located in a bin once the watermark is hided in the signal by using Cho et al.'s watermarking method [37]. A bit "−1" is embedded by decreasing the mean value, and a bit "+1" is embedded by increasing the mean value. Note that the distances are normalized and are illustrated on x and y axis, respectively, based on the distances between the barycenter of the mesh and the probability of the occurrence from the vertices.

In addition, Fig. 4 shows the histogram's variance which is located in a bin once the watermark is hided in the signal by using Cho et al.'s watermarking method [37]. A bit "−1" is embedded by decreasing the variance, and a bit "+1" is embedded by increasing the variance. Note that the distances are normalized and are illustrated on x and y axis, respectively, based on the distances between the center of the mesh gravity and the probability of the occurrence from the vertices.

Two watermarking schemes by using spherical coordinates were presented in [38]. For embedding the watermark, the radial components are modified. For this purpose, initially, the barycenter is computed and the alignment of the principal axis is made by using PCA. The Cartesian coordinate of the model is then transferred to the spherical coordinate system. The results in [37, 38] showed that they tolerate some malicious and non-malicious attacks. Application of spherical coordinate is not limited to aforementioned methods. In another category of watermarking methods including [39, 40], spherical coordinates act as the carriers of watermark. In a category of watermarking methods, geometry modification is utilized for inserting the watermark. This category is famous because of its high robustness; however, some of them have a lower level of robustness in comparison with their contributors. Two blind watermarking methods have been introduced for measuring the local neighborhood in order to find vertices with a low distortion, [41]. Furthermore, local geometric perturbations embed the watermark bits into these vertices.

3D watermarking methods normally use the original 3D object for the purpose of insertion and extraction of watermark bits. Nevertheless, contour information of

Fig. 6.6 A bunny model is represented as (**a**) geodesic map (**b**) iso-geodesic map [17]

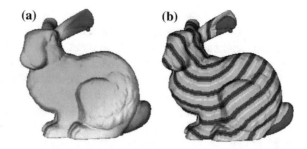

(a)   (b)

the 3D objects was utilized in [42]. Watermarking based on statistical information was introduced in [43]. This method measures the geodesic distances from a certain location on the 3D model and groups them into each region. For the purpose of watermarking, the values of the variance or the mean are modified. The embedding the watermark and the extraction of the watermark are generally similar to [37]. This process is illustrated for a bunny model in Fig. 6.6. It shows the regions comprising the same geodesic distances on the geodesic map.

As depicted in Fig. 6.6, a single reference point is assumed that is highlighted by red color on the bunny's ear, and the geodesic distances are calculated and illustrated with a pseudo-color. Also, the variation of the pseudo-color in the geodesic map from blue to red is based on the geodesic distances. In addition, strip generation of the iso-geodesic map is shown. Almost equal geodesic distances comprise a strip, and one bit is embedded for each strip.

Common operations on the 3D model including smoothing and noise addition are normally tolerated by some watermarking schemes, but mesh editing operations are not easy to resist. Mesh editing operations modify the shape of a 3D mesh model globally and are more realistic than other processing operations.

A resilient watermarking scheme against mesh modeling was presented in [44]. This method embeds the watermark bits by replacing the high values of curvature in the mesh model with the significant mesh patches. As a result of the sensitivity of HVS to the modifications in the regions with low curvature, embedding the watermark bits in these areas achieves the least visible distortion. Since the mesh editing does not considerably change when the geodesic distances are applied, they are preferred to be used instead of the Euclidean distances. This method is not blind, does not tolerate non-uniform scaling and shearing, and needs additional information for the extraction of the watermark.

6.5.2.2 Transform-Based Watermarking

A solution for improving the robustness and imperceptibility of 3D watermarking schemes is achieved by the spectral analysis [6]. This idea was formerly applied in image watermarking. The coefficients are calculated by taking the original image to transform domain and benefit the insertion of the watermark bits to the coefficients in the transform domain, e.g., DFT, DCT, and DWT.

Lack of proper parameters and analysis tools in spectral domain for the decomposition purposes of arbitrary 3D mesh models is still big challenges for achieving a generalized watermarking approach. It becomes worse when arbitrary triangular meshes are encountered [45]. However, similar results to traditional wavelet transforms are achieved through analysis on multi-resolution representation of the mesh surface.

In line with multi-resolution decomposition of 3D models, a watermarking method was proposed in [46] that benefits the edge collapse-based decomposition of Hoppe's method [47] once the scalar basis functions are applied to the original mesh model. These functions are related to the most significant perceptible properties of 3D models and enhance the robustness of the watermarking method. The overall process is similar to the application of the functions like DCT that carries the watermark bits along with its coefficients in large amplitudes. This method is non-blind and needs mesh registration and resampling.

Another non-blind watermarking scheme was introduced in [48] based on a theoretical wavelet analysis method in [49] for surfaces with random topology. This watermarking method modifies the coefficients of wavelet for embedding the watermark bits. An extension of this method for semi-regular 3D mesh models was presented in [50]; however, it is a blind watermarking scheme. An analysis method for wavelet, but on irregular 3D mesh models, was introduced in [51]. It is then utilized in a watermarking scheme that works on both regular and irregular triangular meshes [52]. It modifies the L2-norm of the coefficients for embedding the watermark bits.

Fragile watermarking benefits from multi-resolution analysis of wavelet as presented in [53]. Initially, the wavelet decompositions are applied to the original 3D model. Then, the facets in the coarser mesh are obtained and taken as authentication primitives. The purpose of this method is the authentication of semi-regular meshes. Although it is resilient against similarity transformations, it suffers from low localization and causality problem [45]. By minor modification of norms and orientations of the vectors of wavelet coincident, authenticating watermark was presented in [54]. This method is sensitive to remeshing and geometric alterations whether locally or globally, but it is able to tolerate the content-preserving attacks.

Burt–Adelson pyramid was applied in [55] by using the idea in [56] to design a robust watermarking method. The watermark bits are then embedded properly in the coarser mesh representation. This method is non-blind, as registration and resampling are necessary for the extraction of the watermark bits. As the original location, orientation, scale, topology, and resolution level of the 3D model are altered followed by the attacks, the process of registration and resampling guarantees that the 3D model is restored to its original manner.

There are other frequency-based 3D watermarking methods in the literature. A robust and basic method in this category was introduced in [57]. It assumes a seed vertex in the mesh model. The neighboring vertices to the seed vertex, by considering a predefined topological distance, are gradually added to an area, known as a patch. The predefined distance is obtained from Dijkstra's algorithm. Therefore, the original 3D mesh model is marked by several patches. Each patch contains its spectral coefficients which come from the Kirchhoff matrix [58]. Finally, by additive modulation of the coefficients, watermark bits are embedded into the patches.

The robustness of the method can be boosted through duplication of the watermark bits. Moreover, it successfully tolerates simplification, smoothing, and cropping of the mesh model. This method is categorized as non-blind.

Geometry compression was achieved by using spectral decomposition [16]. It was developed in a 3D watermarking over triangular mesh [59]. They initially modified the spectral coefficients as the outcome of the Laplacian matrix. Then, embedding the watermark bits was performed on the coefficients. Watermarking process through spectral analysis is normally time-consuming and complicated, as it needs the results of computations of the eigenvalues and eigenvectors on large matrices, e.g., Laplacian or Kirchhoff. As a solution, geometry compression in [60] is based on partitioning the original 3D mesh model to a group of submeshes in smaller sizes. Each submesh is compressed by means of spectral compression. The spectral coefficients corresponded to each submesh are lastly modified for embedding the watermark bits. In another attempt, a fast watermarking method [61] benefits the orthogonalization of a small set of radial basis functions for obtaining a less computational complexity. It is suitable for extensively large 3D models and avoids computations of spectral analysis and multi-resolution.

6.6 Distortion Evaluation in 3D Watermarking

When a 3D mesh model carries the watermark, it is important to recognize how the original model and the watermarked model differ and how errors modified the outcome model. Let us consider the original mesh object O and the processed object $\hat{O}$. Six classes of measurements are employed for distortion evaluation. They are Hausdorff distance comparison, volume-based measurement, energy minimization measurement, curvature-based measurement, projection image comparison, and human perceptual distance.

Hausdorff distance is the most applicable measurement [62]. It measures the similarity of simplified 3D image [63] and original 3D image [64]. As the Hausdorff distance is not symmetric, two distances are evaluated including forward $E_f\left(O,\hat{O}\right)$ and backward $E_b\left(\hat{O},O\right)$. The maximum root mean square (MRMS) error between O and $\hat{O}$ is then used for the evaluation of the distortion of watermarking scheme. Metro tool is a well-known tool for computing the MRMS by measuring the surface error between two 3D meshes. Among other distortion evaluation measurements, Hausdorff distance is the most accurate and comprehensive measurement in computer vision and computer graphics. Hausdorff distance is expressed as in Eq. (6.1).

$$E\left(O,\hat{O}\right) = \text{MAX}\left\{ E_f\left(O,\hat{O}\right) = \frac{\sum_{v\in O} \min_{\hat{v}\in\hat{O}}\|v-\hat{v}\|}{|O|}, E_b\left(\widehat{O,O}\right) = \frac{\sum_{\hat{v}\in\hat{O}} \min_{v\in O}\|\hat{v}-v\|}{|O|} \right\}$$

$$(6.1)$$

Volume-based measurement is defined as $V(O, \hat{O})$ and measures the error between O and $\hat{O}$ [65]. V is a Lebesque formula. The volume between two objects is minimum when $\hat{O}$ is the best approximation of O.

Energy minimization measurement measures the sum of squared distances between the two surfaces in form of a scalar value [5]. Those simplification algorithms that utilize the edge collapse strategy are commonly evaluated by the minimum energy. Quadric error metric (QEM) [5] and elastic equations [47] have been used to measure the similarity of the surfaces; however, QEM is more common, accurate, and efficient than elastic equations. They are implicitly related to the other measures, e.g., Hausdorff distance, volume, and curvature.

Curvature-based distance measures the visual distortion with respect to the sensitivity of HVS to the modification in curvature direction [66]. The distance, tangent, and curvature are three main components of a local error after decomposition. The visual similarity between two meshes can be evaluated by the Lavoué's perceptually driven roughness measurement [67] and objective roughness measurement [68].

Projection image comparison measures the similarity between the 2D projection images of 3D meshes and evaluates the images based on the conceptual features of HVS. In other words, it measures the distortion level from the perspective of the human eye [69].

Signal-to-noise ratio (SNR) is a well-known measurement for evaluating the distortion level of 2D signals. It can be defined as the ratio between the sum of the vertex coordinates and the Euclidean distance between vertices. As Euclidean distance is not an effective measurement for distortion between two surfaces, SNR cannot be taken as a suitable measurement for 3D images [38].

Human perceptual distance The human eye naturally measures the similarity of meshes in a subjective manner. On the contrary, objective methods measure the distance between two meshes. The mesh processing distortion assessment based on the human perceptual distance can be found in [70].

References

1. Doerr, G., and J.-L. Dugelay. 2003. A guide tour of video watermarking. *Signal Processing: Image Communication* 18(4): 263–282.
2. Ohbuchi, R., H. Masuda, and M. Aono. 1997. Watermaking three-dimensional polygonal models. In *Proceedings of the fifth ACM international conference on Multimedia*. ACM.
3. Stollnitz, E.J., T.D. DeRose, and D.H. Salesin. 1996. *Wavelets for computer graphics: Theory and applications*. Morgan Kaufmann.
4. Eck, M., et al. 1995. Multiresolution analysis of arbitrary meshes. In *Proceedings of the 22nd annual conference on computer graphics and interactive techniques*. ACM.
5. Garland, M., and P.S. Heckbert. 1997. Surface simplification using quadric error metrics. In *Proceedings of the 24th annual conference on computer graphics and interactive techniques*. ACM Press/Addison-Wesley Publishing Co.
6. Cox, I.J., et al. 1997. Secure spread spectrum watermarking for multimedia. *IEEE Transactions on Image Processing* 6(12): 1673–1687.

7. Watt, A.H., and A. Watt. 2000. *3D computer graphics*, vol. 2. Addison-Wesley Reading.
8. Dugelay, J.-L., A. Baskurt, and M. Daoudi. 2008. *3D object processing: compression, indexing and watermarking*. Hoboken: John Wiley & Sons.
9. Li, H., P. Roivainen, and R. Forchheimer. 1993. 3-D motion estimation in model-based facial image coding. *IEEE Transactions on Pattern Analysis and Machine Intelligence* 15(6): 545–555.
10. Lindstrom, P., et al. 1996. Real-time, continuous level of detail rendering of height fields. In *Proceedings of the 23rd annual conference on computer graphics and interactive techniques*. ACM.
11. Heckbert, P., and M. Garland. 1994. Multiresolution modeling for fast rendering. In *Graphics interface*. Canadian Information Processing Society.
12. Zeki, A.M., and A. Abubakar. 2013. 3D digital watermarking: issues and challenges. In *Proceeding of the international conference on artificial intelligence and computer science (AICS2013). Artificial intelligence and it's application in life*. Bayview, Langkawi, MALAYSIA.
13. Mackinlay, J.D., S.K. Card, and G.G. Robertson. 1990. Rapid controlled movement through a virtual 3D workspace. In *ACM SIGGRAPH computer graphics*. ACM.
14. Kai Wang, G.L. Florence Denis, Atilla Baskurt.
15. Luo, M. 2006. *Robust and blind 3D watermarking*.
16. Taubin, G., T. Zhang, and G. Golub. 1996. *Optimal surface smoothing as filter design*. Berlin: Springer.
17. Yang, Y. 2013. *Information analysis for steganography and steganalysis in 3D polygonal meshes*. Durham University.
18. Ohbuchi, R., H. Masuda, and M. Aono. 1997. *Embedding data in 3D models*. In *Interactive distributed multimedia systems and telecommunication services*. Berlin: Springer.
19. Cayre, F., and B. Macq. 2003. Data hiding on 3-D triangle meshes. *IEEE Transactions on Signal Processing* 51(4): 939–949.
20. Bas, P. 2000. *Méthodes de tatouage d'images fondées sur le contenu*.
21. Wang, C.M., and Y.M. Cheng. 2005. An efficient information hiding algorithm for polygon models. In *Computer graphics forum*. Wiley Online Library.
22. Cheng, Y.-M., and C.-M. Wang. 2007. An adaptive steganographic algorithm for 3D polygonal meshes. *The Visual Computer* 23(9–11): 721–732.
23. Chao, M.-W., et al. 2009. A high capacity 3D steganography algorithm. *IEEE Transactions on Visualization and Computer Graphics* 15(2): 274–284.
24. Mao, X., M. Shiba, and A. Imamiya. 2001. Watermarking 3D geometric models through triangle subdivision. In *Photonics West 2001-electronic imaging*. International Society for Optics and Photonics.
25. Amat, P., et al. 2010. Lossless 3D steganography based on MST and connectivity modification. *Signal Processing: Image Communication* 25(6): 400–412.
26. Cheng, Y.-M., and C.-M. Wang. 2006. A high-capacity steganographic approach for 3D polygonal meshes. *The Visual Computer* 22(9–11): 845–855.
27. Bogomjakov, A., C. Gotsman, and M. Isenburg. 2008. Distortion-free steganography for polygonal meshes. In *Computer graphics forum*. Wiley Online Library.
28. Tu, S.-C., et al. 2010. An improved data hiding approach for polygon meshes. *The Visual Computer* 26(9): 1177–1181.
29. Chou, C.-M., and D.-C. Tseng. 2006. A public fragile watermarking scheme for 3D model authentication. *Computer-Aided Design* 38(11): 1154–1165.
30. Yeo, B.-L., and M.M. Yeung. 1999. Watermarking 3D objects for verification. *IEEE Computer Graphics and Applications* 19(1): 36–45.
31. Lin, H.-Y.S., et al. 2005. Fragile watermarking for authenticating 3-D polygonal meshes. *IEEE Transactions on Multimedia* 7(6): 997–1006.
32. Wu, H.-T., and Y.-M. Cheung. 2005. A fragile watermarking scheme for 3D meshes. In *Proceedings of the 7th workshop on multimedia and security*. ACM.

33. Wang, Y.-P., and S.-M. Hu. 2009. A new watermarking method for 3D models based on integral invariants. *IEEE Transactions on Visualization and Computer Graphics* 15(2): 285–294.
34. Wang, W.-B., et al. 2008. A numerically stable fragile watermarking scheme for authenticating 3D models. *Computer-Aided Design* 40(5): 634–645.
35. Yeung, M., and B.-L. Yeo. *1998.* Fragile watermarking of three-dimensional objects. In *Proceedings. 1998 International Conference on Image processing, ICIP 98*. IEEE.
36. Yu, Z., H.H. Ip, and L. Kwok. 2003. A robust watermarking scheme for 3D triangular mesh models. *Pattern Recognition* 36(11): 2603–2614.
37. Cho, J.-W., R. Prost, and H.-Y. Jung. 2007. An oblivious watermarking for 3-D polygonal meshes using distribution of vertex norms. *IEEE Transactions on Signal Processing* 55(1): 142–155.
38. Zafeiriou, S., A. Tefas, and I. Pitas. 2005. Blind robust watermarking schemes for copyright protection of 3D mesh objects. *IEEE Transactions on Visualization and Computer Graphics* 11(5): 596–607.
39. Ashourian, M., R. Enteshari, and J. Jeon. 2004. Digital watermarking of three-dimensional polygonal models in the spherical coordinate system. In *Proceedings of the Computer graphics international*, 2004. IEEE.
40. Darazi, R., R. Hu, and B. Macq. 2010. Applying spread transform dither modulation for 3D-mesh watermarking by using perceptual models. In *2010 IEEE international conference on acoustics speech and signal processing (ICASSP)*. IEEE.
41. Bors, A.G. 2006. Watermarking mesh-based representations of 3-D objects using local moments. *IEEE Transactions on Image Processing* 15(3): 687–701.
42. Wang, X., W. Qi, and P. Niu. 2007. A new adaptive digital audio watermarking based on support vector regression. *IEEE Transactions on Audio, Speech, and Language Processing* 15(8): 2270–2277.
43. Bennour, J., and J.-L. Dugela. 2006. Protection of 3D object visual representations. In *IEEE international conference on multimedia and expo*. IEEE.
44. Lin, C.-H., et al. 2010. A novel semi-blind-and-semi-reversible robust watermarking scheme for 3D polygonal models. *The Visual Computer* 26(6–8): 1101–1111.
45. Wang, K., et al. 2008. A comprehensive survey on three-dimensional mesh watermarking. *IEEE Transactions on Multimedia* 10(8): 1513–1527.
46. Praun, E., H. Hoppe, and A. Finkelstein. 1999. Robust mesh watermarking. In *Proceedings of the 26th annual conference on computer graphics and interactive techniques*. ACM Press/Addison-Wesley Publishing Co.
47. Hoppe, H. 1997. *View-dependent refinement of progressive meshes*. In *Proceedings of the 24th annual conference on computer graphics and interactive techniques*. ACM Press/Addison-Wesley Publishing Co.
48. Kanai, S., H. Date, and T. Kishinami. 1998. Digital watermarking for 3D polygons using multiresolution wavelet decomposition. In *Proceedings of the sixth IFIP WG*. Citeseer.
49. Lounsbery, M., T.D. DeRose, and J. Warren. 1997. Multiresolution analysis for surfaces of arbitrary topological type. *ACM Transactions on Graphics (TOG)* 16(1): 34–73.
50. Uccheddu, F., M. Corsini, and M. Barni. 2004. Wavelet-based blind watermarking of 3D models. In *Proceedings of the 2004 workshop on multimedia and security*. ACM.
51. Valette, S., and R. Prost. 2004. Wavelet-based multiresolution analysis of irregular surface meshes. *IEEE Transactions on Visualization and Computer Graphics* 10(2): 113–122.
52. Kim, M.-S., et al. 2005. Watermarking of 3D irregular meshes based on wavelet multiresolution analysis. In *Digital Watermarking*, 313–324. Berlin: Springer.
53. Cho, W.-H., et al. 2004. Watermarking technique for authentication of 3-D polygonal meshes. In *Digital Watermarking*, 259–270. Berlin: Springer.
54. Wang, K., et al. 2008. A fragile watermarking scheme for authentication of semi-regular meshes. *Proceedings of the Eurographics Short Papers* 8(5–8): 36.
55. Yin, K., et al. 2001. Robust mesh watermarking based on multiresolution processing. *Computers and Graphics* 25(3): 409–420.

56. Guskov, I., W. Sweldens, and P. Schröder. 1999. Multiresolution signal processing for meshes. In *Proceedings of the 26th annual conference on computer graphics and interactive techniques*. ACM Press/Addison-Wesley Publishing Co.

57. Ohbuchi, R., A. Mukaiyama, and S. Takahashi. 2002. A frequency-domain approach to watermarking 3D shapes. In *Computer graphics forum*. Hoboken: Wiley-Blackwell.

58. Bollobás, B. 2013. *Modern graph theory*, vol. 184. Berlin: Springer Science & Business Media.

59. Cayre, F., et al. 2003. Application of spectral decomposition to compression and watermarking of 3D triangle mesh geometry. *Signal Processing: Image Communication* 18(4): 309–319.

60. Abdallah, E.E., A.B. Hamza, and P. Bhattacharya. 2007. Spectral graph-theoretic approach to 3D mesh watermarking. In *Proceedings of graphics interface 2007*. ACM.

61. Wu, J., and L. Kobbelt. 2005. Efficient spectral watermarking of large meshes with orthogonal basis functions. *The Visual Computer* 21(8–10): 848–857.

62. Gelasca, E.D., et al. 2005. Objective evaluation of the perceptual quality of 3D watermarking. In *IEEE international conference on image processing, ICIP 2005*. IEEE.

63. Huttenlocher, D.P., G.A. Klanderman, and W.J. Rucklidge. 1993. Comparing images using the Hausdorff distance. *IEEE Transactions on Pattern Analysis and Machine Intelligence* 15(9): 850–863.

64. Guezlec, A. 2001. "Meshsweeper": dynamic point-to-polygonal mesh distance and applications. *IEEE Transactions on Visualization and Computer Graphics* 7(1): 47–61.

65. Alliez, P., et al. 1999. Mesh approximation using a volume-based metric. In *Proceedings of the seventh pacific conference on computer graphics and applications*. IEEE.

66. Kim, S.-J., S.-K. Kim, and C.-H. Kim. 2002. Discrete differential error metric for surface simplification. In *Proceedings of the 10th pacific conference on computer graphics and applications*. IEEE.

67. Lavoué, G. 2009. A local roughness measure for 3D meshes and its application to visual masking. *ACM Transactions on Applied perception (TAP)* 5(4): 21.

68. Corsini, M., et al. 2007. Watermarked 3-D mesh quality assessment. *IEEE Transactions on Multimedia* 9(2): 247–256.

69. Reddy, M. 2001. Perceptually optimized 3D graphics. *IEEE Computer Graphics and Applications* 5: 68–75.

70. Bian, Z., S.-M. Hu, and R.R. Martin. 2009. Evaluation for small visual difference between conforming meshes on strain field. *Journal of Computer Science and Technology* 24(1): 65–75.

Part III
Document Watermarking

Chapter 7
Natural Language Watermarking

7.1 Introduction

Nowadays, a mass traffic of Internet is occupied by text data transactions. Because text data is widely distributed, searched, and reused in various applications, it is essential to control the copyright over text as well as other forms of data including video, image, and audio. Semantic and syntactic structures of text are good candidates for embedding watermarks. This kind of watermark is known as natural language (NL) watermarking. In multimedia watermarking, the perceptual limitations of HVS and HAS as well as redundancy have widely utilized for watermarking of 2D and 3D signals. On the contrary, NL watermarking cannot freely utilize redundancy of data for embedding watermark. It is due to syntactical and discrete nature of NL that is corresponded to two main features as follows:

1. Combination of syntax and semantics forms both syntactic and semantic structures of a NL content.
2. Combinatorial syntax defines both syntactic and semantic structures of an NL content. It can causally result the sensitivity to operate sentences.

In this chapter, firstly, major concepts and terminologies of natural language processing (NLP) are described. Secondly, the related tools for semantic analyzer, parsers, and generators are explained. Finally, different NL watermarking techniques are discussed.

7.2 Background of Natural Language Processing

Since NL watermarking is just a narrow part of NLP science, it is necessary to be familiar with basic concepts and terminologies in NLP. On the other words, understanding techniques and trends in NL watermarking without studying some prerequisite

© Springer Science+Business Media Singapore 2017
M.A. Nematollahi et al., *Digital Watermarking*, Springer Topics
in Signal Processing 11, DOI 10.1007/978-981-10-2095-7_7

in NLP is not possible. For this purpose, a comprehensive definition on NLP concepts and terminologies is provided in this section which can be found in [1, 2].

Finding automatic methods for analysis, and understanding and generating NL are the main goals in NLP. In the following section, some basic concepts in NL watermarking including data resources, linguistic transformations, NL parsing, word sense disambiguation, statistical language models, NL generation, and NL paraphrasing are explained briefly.

7.2.1 Data Resources

Achievement of hiding information in an effective manner does not come true unless some suitable models can be found for cover medium. This aim is obtained by using large datasets. One of the statistical samples of NL for texts is called as a corpus. The content of a corpus needs to be marked with metadata for better usage. Normally, the outcome of statistical processes and analysis on a corpus is in the form of mass data (corpora). This volume of corpora should be created in electronic formats to be readable by machines for training the NLP models and benchmarking as many of them are available from Linguistic Data Consortium [3]. Table 7.1 presents the major corpora in NLP.

Corpora are not the sole means to access large data related to NLP. There are large databases that store the lexical relations between words in the form of electronic dictionaries, e.g., WordNet [4]. WordNet keeps the nouns, verbs, adjectives, and adverbs of English language along with their synonym set, when each set represents an underlying lexical concept.

7.2.2 Linguistic Transformations

For any natural language, a systematic method is required. When a systematic method is available, it is simple to modify or transform the content of a given text

Table 7.1 Available corpora in linguistic data consortium

Name of corpus	Number of words	Details
Penn Treebank II	1,000,000	Printed 1989
Reuters	810,000,000	Collected between 1996 and 1993. British English
Wall Street Journal	40,000,000	Collected between 1987 and 1993 American English
Susanne	130,000	Subset of Brown corpus
Lanchester-Oslo-Bergen	1,000,000	British English Subset of Brown corpus
Brown	1,000,000	15 different types Printed 1961 American English

in such a way that the concept of the text remains intact and the readers could not recognize the modifications on the text. This is the idea behind NL watermarking. Ideally, the transformations should not change the grammaticality of the sentences. This increases the robustness of watermarking when statistical attacks occur.

The modification on the NL is generally performed by three linguistic transformations, including lexical transformations [5], syntactic transformations [6], and semantic transformations [7]. Lexical transformation that normally embeds the watermark by replacing the word with the synonyms is the commonest NL watermarking method. WordNet dictionary is popular in lexical transformation; however, for acceptable and imperceptible modification of the meaning of the sentences, human supervision is essential [8]. As an instance, there are 10 different synonyms for the word "bank" in WordNet, e.g., depository financial institution, sloping land, and a flight maneuver. The selection of the proper word with the correct sense is a challenge in NLP as is known as the word sense disambiguation task [9].

Syntactic transformations modify the syntax of the sentences on the condition that the least possible alteration of the meaning of the sentence occurs, e.g., passivation and clefting [10]. Various methods of watermarking in NL utilize syntactic transformations [6]. Table 7.2 presents some on the commonest syntactic transformations in English.

Semantic transformations modify the semantic relations among the words or the meaning structure of the sentences, although by using some methods the meaning of the sentence can be preserved. A solution for this purpose is using noun phrase coreferences [7] that refer to the same entity. A transformation method based on the coreference concept is coreferent pruning which deletes the duplicated information about the coreference. On the other hand, the coreferent grafting duplicates the information about a coreference in other sentences or adds additional information on the NL by means of a fact database. Semantic transformation examples were explained in [11].

Table 7.2 Syntactic transformation in English

Name of the transformation	Initial sentence	Transformed sentence
Fronting	*Carefully*, he removed the lid	He removed the lid *carefully*
Pronominalization	I go to the kitchen	I go there
There-construction	A cat is in the home	There is a cat in the home
Preposing	She bought *a pair of gloves with silk embroidery*	*A pair of gloves with silk embroidery* is what she bought
Extraposition	Someone *who we don't know* left a message	Someone left a message *who we don't know*
Clefting	She bought *a pair of gloves with silk embroidery*.	It was *a pair of gloves with silk embroidery* that she bought
Topicalization	I am terrified of *those dogs*	*Those dogs*, I am terrified of
Passivation	Suzan teaches Julie	Julie is taught by Suzan

The tough part in NLP is finding accurate coreference resolution to be performed in semantic transformation [12]. However, finding two coreferent phrases with the same connotations is not always possible. Substitution of such phrases normally yields to the sentences with severe modifications in their semantic structure, e.g., in "Spiderman just saved us from death" and "Peter Parker just saved us from death" the phrases "Spiderman" and "Peter Parker" show this situation.

7.2.3 Natural Language Parsing

Natural language parsing processes and generates a structure for the input sentences [13]. The output of this process can be a semantic, syntactic, or morphologic structure of the sentence or even a combination of them. The output structure assists to obtain the roles of the constituent words in the structure. A parser starts from a given part of the speech. The words are initially classified based on their roles (e.g., noun, adjective, or verb) or are break up to their morphemes by using a morphological analyzer. The output of the parser is in a transformed form of the sentence and a dependency tree for each sentence. It is presented in Fig. 7.1 for the sentence "I am trying to compute the relationship."

The action of parsers is generally similar to image watermarking when the time and the frequency domains as the representations of image are compared to the input NL and the tree relationships of the parser. For simple generation of

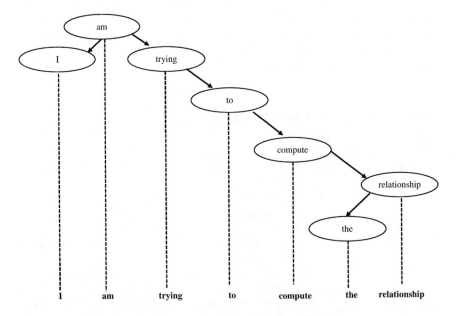

Fig. 7.1 Dependency tree for sentence "I am trying to compute the relationship"

dependency trees, some tools have been introduced in some studies for illustration of the argument or modifier relations [14]. Some of the common parsers include XTAG, Charniak, Collins, and Link [15].

The parser tree (syntactic tree) are using some abbreviations to represent a sentence. In the following, some of these abbreviations are explained.

- *S* is the highest level in tree and is equal to root node of the tree.
- *NP* is corresponding to a noun phrase.
- *VP* is corresponding to a verb phrase and it uses for predication purposes.
- *V* is a verb.
- *N* is a noun.
- *D* is a determiner like "the".

7.2.4 Word Sense Disambiguation

When a word has several meaning, it is known as ambiguous. The aim of word sense disambiguation is resolving the problem of ambiguous words. A more specific linguistic term for ambiguous words is homograph. When several words have the same spelling, but several meaning, origins, and even pronunciations, they are homographs. As an instance, "bank" is a homograph, because it can refer to a financial institution, a slope in the turn of a road, or the edge of a stream. Finding the most accurate and the correct meaning of a word from the given context is a complicated process in NLP [16] and is the main challenge of the word sense disambiguation task.

7.2.5 Statistical Language Models

Statistical language model (LM) is the science of finding the probability of a word occurrence in a NL according to the words previously used in the NL. In other words, probabilities of n-gram word strings can be estimated by LM [17]. The main application of LM is to generate robust watermarking systems that utilize patterns for unmodified and steganographically modified NL [5, 18].

As the occurrence of a word depends on the n-1 preceding word, LM is modeled by Markov models where an n-gram model is an $n - 1$th order Markov model. Considering W as the probability of a set of n consecutive word; where $W = \{w_1, \ldots, w_n\}$ and the initial condition $P(w_1|w_0)$ as a suitable value, the LM probability obtained from Eq. (7.1) is as follows.

$$P(w_1, \ldots, w_n) = \prod_{i=1}^{n} P(w_i|w_0, \ldots, w_{i-1}) \tag{7.1}$$

There is another common measure in signal processing that works based on quantity rather than the model entropy. It is known as perplexity and represents the goodness of fit of a LM for a given text data. The perplexity of a given LM, labeled as L, is obtained by Eq. (7.2) as follows.

$$\text{perplexity}(\iota) = 2^{-\frac{1}{N}\sum \log_2^{P(\text{data}|\text{model})}} \tag{7.2}$$

7.2.6 Natural Language Generation

An attractive task in NLP is converting non-linguistic information representations to natural language output which is known as natural language generation (NLG). NLG needs some communication specifications and maps meaning to text. NLG requires three phases before completion including discourse planning, sentence planning, and surface realization [19]. The process of information selection and the organization of it into coherent paragraphs are performed by discourse planning. Proper selection of words and structures to fit the information into sentence-sized units is the mission of sentence planning. At last, the output format (the surface) is determined by surface realization. The word order, morphology, and final formatting or intonation are finalized in this phase.

According to three phases of NLG, the main components of a NLG system are Including communicative goal, knowledge base, discourse planner, discourse specification, surface realizer, and NL output. It should be mentioned that knowledge base is formed discourse planner. Also, discourse planner in turn is formed discourse specification which is fed to surface realizer. Finally, NL output is created. The nature of a NLG system's inputs relies on the domain specified by the NLG tool. One of the famous weather forecasting systems is the Forecast Generator (FOG) [20]. FOG is a comprehensive NLG system able to produce bilingual NL in English and French. FOG is fed by raw meteorological data and forecasts the weather as the output. Some instances of other NLG systems can be found in [21].

Extra information is normally added to the original text by NL watermarking. It can be performed by altering the structural representation of the NL. For a good data hiding purpose, when text is hided among NL information, NLG system acts as the key role to convert back the watermarked text into NL. NLG can also cover the text generation mechanisms in natural language steganography systems.

7.2.7 Natural Language Paraphrasing

Natural language paraphrasing is a well-known process in NL watermarking. The paraphrasing refers to the process of NL modification or rewriting the text, in such a way that the result be a NL with different length, readability, and style, without losing the original meaning of the NL.

The language of the origin and the designated texts are the same in paraphrasing that is the key difference between text paraphrasing and machine translation.

Paraphrasing systems semantically collect or create sets of equivalent words, phrases, and patterns for their purpose.

Paraphrasing systems are mainly based on creating or collecting sets or pairs of semantically equivalent words, phrases, and patterns. The following sentences from a news story are taken into account [22]:

"After the latest Fed rate cut, stocks rose across the board."

"Winners strongly outpaced losers after Greenspan cut interest rates."

As understood, both sentences represent the same and are counted as a semantically related pair. A paraphrasing system should be trained well by multiple sample pairs, prior to operate in real applications.

7.3 Natural Language Watermarking

Watermarking systems face with the problem of managing the right protection. Digital signatures are common for protection of natural language texts, but they are repeatable from the original text [23]. Fingerprinting and duplicate detection are other techniques for protection of data [24]. Some watermarking techniques insert additional information to the original data, e.g., metadata, except for the case of carrying 1-bit watermarks. Duplicate detection identifies the duplicates only with no detection of the first owner of the document. Therefore, they are good for finding the plagiarism but where ability to assert priority is essential using them becomes a challenge of NL watermarking.

So far, NLP was discussed along with its method and characteristics. In the reminder of this chapter, NL watermarking which is almost a new research area and needs comprehensive discussions is discussed. The general classes of NLP watermarking are brought in Fig. 7.2.

In the following, the major requirements and techniques for NL watermarking are discussed.

7.3.1 Requirements of NL Watermarking

The performance of NL watermarking is measured based on the trade-off between multiple criteria. The robustness (the preserve-ability of the text value against adversary attacks), capacity (the length of the embedded watermark bits in a document), and stealthiness (imperceptibility of the watermark for different characteristics of document (genre), reader, and writer) are the major comparing criteria. NL watermark techniques should observe other factors including meaning, grammaticality, and style. The most important factor is the meaning. The meaning (the value) of a NL must be kept intact during the watermarking. The meaning is

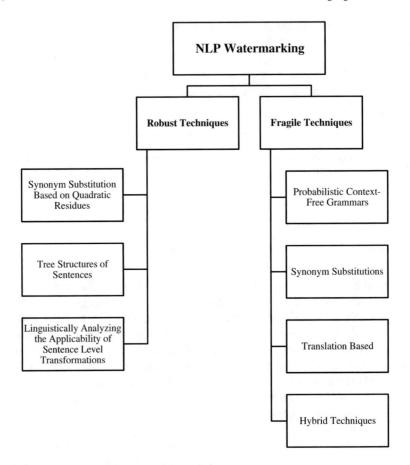

Fig. 7.2 An overview of NL watermarking techniques

subjected to further modifications by grammaticality and fluency. The watermarking technique must guarantee that the fluency and grammar rules of NL are not violated. According to the text genre and literature, each author has his signature style of writing. Indeed, watermarking must have considerations on preserving the author's style.

A high-capacity NL watermarking technique hides the highest possible number of watermark bits in a given document. The capacity of watermarking is limited by some constraints. Sometimes transformation cannot be applied to a given word, or finding alternative words for a given word or sentence is not always feasible as a result of grammatical and vocabulary constraints. Moreover, the reversibility of linguistic transformation makes NL watermarking susceptible to removal attacks in sentence level.

As discussed in previous sections, the parsing and part-of-speech-tagging tools are good to perform their isolated tasks; however, they were not designed for embedding the watermark. Consider a natural language generation system that performs text-to-text generation to convert and rewrite the text in a robust manner. For this purpose, an application-specific model is utilized [25]. RealPro is a natural language generation system that uses a specific representation for sentence structure (i.e., DsuntS) as the system's input. This system can test the watermarking in sentence level, as well.

In general, for embedding a message into a given cover document as used as the basis of linguistic NL watermarking systems [18, 26], five steps are performed. They are linguistic analysis, selection, embedding, generating the surface form, and verification of embedding. The general structure of a NL watermarking system is illustrated in Fig. 7.3.

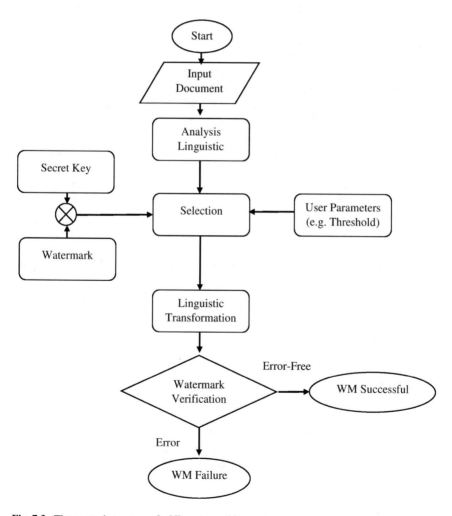

Fig. 7.3 The general structure of a NL watermarking system

Initially, the syntactic and semantic structure of the document that is given for embedding the watermark should be analyzed. In linguistic analysis, some operations are usually performed on the original document, e.g., either part-of-speech tagging [27], syntactic parsing [26], or semantic analysis [7]. In some watermarking systems, the analysis system needs training corpus. This process acts as a secret key that is separately selected for information hiding in every session [28]. After analysis of the structure of the document, the candidate words or sentences for embedding the watermark are chosen. For this purpose, the author usually set the user parameters in such a way that the candidates with the highest potentiality of embedding the watermark and the least negative effect on the value of the text, are chosen (e.g., the style of the author). In some cases, the style of the sentences is not very important, but the order of the sentences should be kept fixed, e.g., a text related to a meal recipe. In this step, the system intakes the encrypted watermark message, the secret key, and the user parameters as the input. Then, the embedding step is performed and the linguistic information is embedded into the selected parts of the document.

After embedding the watermark information, the deep structure should be converted into surface form of natural language. In some cases, NLG system is necessary in this step [29]. As an instance when the sentence "this frank discussion will close this chapter" is replaced with "this chapter will be closed by this frank discussion," NLG system is needed. There are other cases that do not require NLG. When the sentence "In seconds, Greg made his decision.", just by reordering the words, is replaced with "Greg made his decision in seconds.", NLG is not necessary [10]. Generally, NLG is neglected when the embedding is performed either by reordering the words, removing some words, or addition of new words to the sentence with no further modifications in the structure of the sentence.

Finally, the new text is analyzed to ensure that the entire embedding process has been performed successfully. Once an embedding is operated, the verification of embedding is performed. Depending on the system design, there are two processes for verification.

(A) The sentence level watermarking: Feature extraction and verification of the features to ensure that the values are properly selected and embedded are required in the sentence level; e.g., it is a passive sentence and carries more than one preposition [29].

(B) The synonym-based watermarking: The output text is checked to ensure that the selected words are successfully replaced with their synonyms and the exact mixed radix code required by the secret message (e.g., the word "wonderful" is replaced by "fine") is carried by the synonyms. This process has already been used in T-Lex system [30]. Usually, there are more than one word to be used as the synonym of a word. The author of the document may assist the automatic verification process by choosing the most suitable word when there are more than one synonym for a given word and by rewriting the sentence or replacing the words when there is no applicable transformation for a word or a sentence.

7.3.2 Natural Language Watermarking Techniques

Fragile watermarking and robust watermarking techniques have been widely used for NL watermarking. Due to difficulty of embedding robust watermarking to the text, there are only few works in the literature for robust NL watermarking. The remainder of this section discusses the fragile NL watermarking techniques and then goes toward the robust NL watermarking techniques.

7.3.2.1 Fragile Natural Language Watermarking Techniques

The main fragile NL watermarking techniques including probabilistic context-free grammars, synonym substitutions, translation based, and hybrid techniques are described in this part.

Probabilistic Context-Free Grammars

A probabilistic context-free grammar (PCFG) is the commonest model for transforming the NL by using context-free grammar rules where each grammar rule has its own probability [1]. It starts from root node and generates strings by rewriting the text recursively, according to a random rule. The reverse of this process is performed by parsing the text to find out the rules have been used for generating the text. This method was slightly modified to become resilient to statistical attacks.

The statistical features of the PCFG's generated string can become near to normal text by the mimicry text approach [31]. For this purpose, a Huffman code is assigned to each grammar rule according to its probability. The PCFG and the probabilities of the grammar rules are come from a corpus. Finally, a grammar rule with the corresponding code to a given portion of the message is selected for embedding the payload string. Ideally, this model is expected to generate meaningful texts, with several sentences, that are close to natural language. In fact, the PCFG-generated strings have a limited coverage and are not very sensual to human readers. Due to this limitation, the output of PCFG is generally transferred on the communication channels and is read by computers.

Synonym Substitutions

NL watermarking methods replace a subset of the words from the document with the proper synonyms. An instance of this method was utilized in T-Lex system [30] where the synonyms are derived from WordNet. A set of synonyms for each word is ordered alphabetically. The following sample shows the sets of synonyms for two candidate words [32].

$$\text{Midshire is a}\begin{Bmatrix} 0 & \text{wonderful} \\ 1 & \text{decent} \\ 2 & \text{fine} \\ 3 & \text{great} \\ 4 & \text{nice} \end{Bmatrix}\text{little}\begin{Bmatrix} 0 & \text{city} \\ 1 & \text{town} \end{Bmatrix}$$

Consider the current string is $(101)2 = 5$ which is translated in mixed radix form as follows.

$$\begin{pmatrix} a_1 & a_0 \\ 5 & 2 \end{pmatrix} = 2a_1 + a_0 = 5$$

where $0 \leq a_1 < 5$ and $0 \leq a_0 < 2$. If the obtained values become $a_1 = 2$ and $a_0 = 1$, then the words "fine" and "town" are selected for generating the string.

For better understanding, consider the sentence "… I can tell you, to be making new acquaintances every day…" from Jane Austen's novel Pride and Prejudice. By using the T-Lex system, this sentence is modified to "… I can tell you, to be fashioning new acquaintances every day…" after embedding the watermark. In the same way, the sentences are as follows:

"An invitation to dinner was soon afterwards dispatched","… and make it still better, and say nothing of the bad–belongs to you alone.", "Bingley likes your sister undoubtedly" after embedding the watermark are modified to:

"An invitation to dinner was soon subsequently dispatched", "… and make it still better, and say nada of the bad–belongs to you alone.", "Bingley likes your sister doubtless.", respectively. These examples prove that the alternative synonyms used by T-Lex system are not always correct in English, as seen in the phrase "soon subsequently dispatched." Moreover, the alternative word does not always follow the genre and the style of the author; e.g., the word "nada" does not fit to Jane Austen's style. It is not even appropriate in the English literature writing. The reason for both problems is the T-Lex system's disregarding of some important factors, e.g., genre, author style, and sentence context, when picking an alternative word from subsets. This problem is seen in all synonym substitution methods. One solution to detect the improper synonyms for a typical style is to train the language models by a collection of typical text with the same genre and style. By using the information derived from language models during the embedding process, many problems associated with improper selection of synonyms can be resolved; however, this process imposes a high computational complexity to the system.

Translation Based

The idea behind translation-based watermarking techniques is using the machine noises on the translation of natural language documents [28]. When a machine automatically translates a document, frequent errors normally occur in the translated text. Therefore, addition of watermark data to the document is seen as the errors and is counted as a part of normal noise added by legitimate automatic text translator. Thus, the reveal of watermark gets harder.

An instance is LiT system [28] that encodes the information by using a keyed hash of translated sentences. LiT system employs multiple machine translation systems to obtain several versions of a document translation. However, for an attacker is possible to reveal the watermark information by statistical analysis of the language models as used by machine translation systems.

Hybrid Techniques

The NICETEXT [33, 34] is a hybrid watermarking system that combines the PCFG and the lexical transformations to achieve a natural cover text. It benefits a large dictionary table and a style template for generating the cover text. Dictionary table is derived from a part-of-speech tagger or WordNet. Each field of the dictionary table is a pair of type or word. For each word, the type is assigned based on the part-of-speech [34] of the word or its synonym set [33]. The idea of PCFG system is presented by style template. The style template improves the generation of natural sequences of part-of-speech by controlling the word generation, punctuation, capitalization, and white spaces. Style templates can be learned and employed from online corpora, e.g., Aesop's Fables and Federal Reserve Board meeting minutes.

7.3.2.2 Robust Natural Language Watermarking Techniques

The first robust NL watermarking techniques utilize linguistic transformations for implementation of the watermarking system. In average 4.5 publications per year [35] have been appeared in the area of Robust NL watermarking techniques since it was firstly introduced in 2000 [36].

Synonym Substitution Based on Quadratic Residues

The robust watermarking based on the synonym substitution was introduced in [37]. This technique benefits ASCII values of the words. A cover text with ASCII value is presented by $A(w)$ where w_i is the current word. Consider k is the number of bits in the watermark message, p as a 20 digit prime key, and $r_0,...,r_{k-1}$ as a sequence of pseudo-random numbers generated by p as seed. The watermark bits are defined as m_i mod k. If m_i mod $k = 1$ and $A(w_i) + r_i$ mod k is a quadratic residue modulo p, then w_i is kept same. Else, it is changed. This method is low robust because an intruder can use synonym substitution and replace his/her watermark message with the document's watermark or scramble the watermark message to become uncover able.

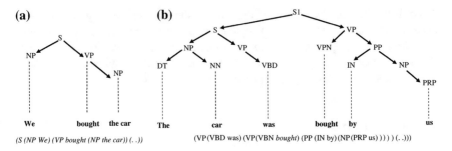

Fig. 7.4 NL watermarking by syntactic tree structure: **a** original sentence; **b** watermarked sentence

Tree Structures of Sentences

The tree-based robust watermarking techniques alter the deep structure of the sentence for embedding the watermark [7, 26]. Despites the lexical substitution techniques, rather embedding the watermark directly to the surface properties of the NL, tree-based robust watermarking techniques hide the watermark over the parsed presentation of the sentence. It makes the modification and the overwrite of the watermark harder which yields to a more robust technique.

Two tree-based techniques have been presented in [7, 26]. The first technique alters the syntactic parse tree for hiding the watermark over the cover text, while the semantic tree representations are modified in the second technique. The representation of the semantic tree structure is a flat text meaning representation of a sentence [7]. As an instance, syntactic tree for the original sentence "We bought the car." and the corresponding watermarked cover text, i.e., "The car was bought by us." is illustrated in Fig. 7.4.

Ontological semantics resources may generate this form of representation of the sentences, as well [38]. When the tree structure of a sentence is formed, some sentences are selected as the carriers of watermark. Initially, the tree structure T_i is formed for a sentence S_i. Then, the nodes of the tree are labeled in preorder traversal of T_i. Considering the one-way hash function $H()$ and p as the secret key, if $j + H(p)$ is a quadratic residue modulo p, the node is converted to 1; otherwise to 0. Following a post-order traversal, a binary sequence B_i is generated for a sentence S_i. By using the equation $d_i = H(B_i)$ XOR $H(p)$, the sentences are ordered based on their rank, d_i. Then, the embedding of the watermark is started from the successor of the sentence s_j, that is, the sentence with the lowest rank. As the sentence s_j is the marker sentence, it points to a sentence that carries the watermark. The watermarking continues by the sentence with the next lowest rank. The watermark bits are applying by a syntactic or semantic transformation once the mark carrying sentences are ranked.

The above-mentioned techniques have the following difficulties. For both selection and embedding the watermark, only one feature of the sentence is applied and a sentence cannot be used for both. Therefore, a sentence acts as the mark carrying and the next immediate sentence becomes the watermark message carrier. It degrades the capacity of watermarking and increases the problem of reordering the sentences, insertion of new sentences, and selection of a subset of sentences. The serious consequence of proximity is the probability of damaging the embedded bit by a random change in any sentence, i.e., $|M|/n$ which can be neglected when texts become large (i.e., $n \gg |M|$). Moreover, the wide application of mentioned techniques relies on semantic parsing and coreference resolution of the sentences in a fully automated manner, while the on-the-shelf natural language processing technology only provides this condition for very domain-specific applications.

Linguistically Analyzing the Applicability of Sentence Level
Transformations

A technique based on the sentence level based on syntactic watermarking was presented in [10]. Several syntactic transformations in sentence level have been linguistically analyzed on 6000 sentences of a hand-parsed corpus [39]. The operations comprise adjunct movement, adjective reordering, and verb particle movement. A detailed analysis on the coverage and the applicability of sentences for watermarking, in addition to the involved challenges in the writing process of a generic transformation rule for a natural language, have been performed in [10]. This technique converts the reordering transformation of the words in a sentence such as adjunct movement, or addition (e.g., clefting) or removing (e.g., that/who be removal) a fixed structure to/from a sentence. As a surface level generator is not utilized by this technique, the number of transformations used for the purpose of analysis is limited.

7.4 NL Watermarking Versus Text Watermarking

NL watermarking and text watermarking differ from multiple points of view. Text watermarking alters the text appearance by changing the words, characters, and lines. On the other hand, NL watermarking modifies the structure and constituent of a sentence to embed the watermark. Due to their differences, different robustness is expected to be achieved by them.

A successful text watermarking technique needs to be optically imperceptible. If the added inter-letter of inter-word spaces can be visually delineated by HVS, the text watermarking is divulged. Moreover, text watermarking techniques are susceptible to reformatting and scanning attacks and optical character recognition (OCR). Therefore, they suffer from a low robustness. From the other side, NL watermarking is widely applied in tamper-proofing, text auditing, traitor tracing, and metadata binding. Moreover, it embeds the watermark data wish a high level of robustness.

References

1. Manning, C.D., and H. Schütze. 1999. *Foundations of statistical natural language processing*, vol. 999. MIT Press.
2. Topkara, M.K. 2007. *New designs for improving the efficiency and resilience of natural language watermarking*. ProQuest.
3. *Linguistic Data Consortium*.
4. Miller, G., and C. Fellbaum. 1998. *Wordnet: An electronic lexical database*. MIT Press Cambridge.
5. Topkara, M., U. Topkara, and M.J. Atallah. 2007. Information hiding through errors: a confusing approach. In *Proceedings of the SPIE international conference on security, steganography, and watermarking of multimedia contents*.
6. Meral, H.M., et al. 2007. *Syntactic tools for text watermarking*. In International society for optics and photonics *electronic imaging 2007*.
7. Atallah, M.J., et al. 2002. Natural language watermarking and tamperproofing. In *Information hiding*. Springer.
8. Bergmair, R., and S. Katzenbeisser. 2004. Towards human interactive proofs in the text-domain. In *Information security*, 257–267. Springer.
9. Ide, N., and J. Vronis. 1998. Word sense disambiguation: the current state of the art. *Computational Linguistics*, 24(1).
10. Murphy, B. 2001. *Syntactic information hiding in plain text*. Trinity College.
11. Singh, R., and S. Gulwani. 2012. Learning semantic string transformations from examples. *Proceedings of the VLDB Endowment* 5(8): 740–751.
12. Ng, V., and C. Cardie. 2002. Improving machine learning approaches to coreference resolution. In *Proceedings of the 40th annual meeting on association for computational linguistics*. Association for Computational Linguistics.
13. Jurafsky, D. 2000. *Speech & language processing*. Pearson Education India.
14. Xia, F., and M. Palmer. 2001. Converting dependency structures to phrase structures. In *Proceedings of the first international conference on Human language technology research*. Association for Computational Linguistics.
15. Manning, D.K.C.D. 2003. Natural language parsing. In *Advances in neural information processing systems 15: proceedings of the 2002 conference*. MIT Press.
16. Yuan, D., et al. 2016. *Word sense disambiguation with neural language models*. arXiv preprint arXiv:1603.07012.
17. Stolcke, A. 2002. *SRILM-an extensible language modeling toolkit*. In *INTERSPEECH*.
18. Taskiran, C.M., et al. 2006. Attacks on lexical natural language steganography systems. In *Electronic Imaging 2006*. International Society for Optics and Photonics.
19. Reiter, E., R. Dale, and Z. Feng. 2000. *Building natural language generation systems*. Vol. 33. MIT Press.
20. Bourbeau, L., et al. 1990. Bilingual generation of weather forecasts in an operations environment. In *Proceedings of the 13th conference on Computational linguistics*, vol 3. Association for Computational Linguistics.
21. Reiter, E. 2010. Natural language generation. *The handbook of computational linguistics and natural language processing*, 574–598.
22. Barzilay, R., and L. Lee. 2003. Learning to paraphrase: An unsupervised approach using multiple-sequence alignment. In *Proceedings of the 2003 conference of the North American chapter of the association for computational linguistics on human language technology*, Vol. 1. Association for Computational Linguistics.
23. Rivest, R.L., A. Shamir, and L. Adleman. 1978. A method for obtaining digital signatures and public-key cryptosystems. *Communications of the ACM* 21(2): 120–126.
24. Schleimer, S., D.S. Wilkerson, and A. Aiken. 2003. Winnowing: local algorithms for document fingerprinting. In *Proceedings of the 2003 ACM SIGMOD international conference on Management of data*. ACM.

25. Soricut, R., and D. Marcu. Stochastic language generation using WIDL-expressions and its application in machine translation and summarization. In *Proceedings of the 21st international conference on computational linguistics and the 44th annual meeting of the Association for Computational Linguistics*. Association for Computational Linguistics.

26. Atallah, M.J., et al. 2001. Natural language watermarking: Design, analysis, and a proof-of-concept implementation. In *Information Hiding*. Springer.

27. Murphy, B., and C. Vogel. 2007. The syntax of concealment: Reliable methods for plain text information hiding. In *Electronic Imaging 2007*. International Society for Optics and Photonics.

28. Stutsman, R., et al. 2006. Lost in just the translation. In *Proceedings of the 2006 ACM symposium on applied computing*. ACM.

29. Topkara, M., U. Topkara, and M.J. Atallah. 2006. Words are not enough: sentence level natural language watermarking. In *Proceedings of the 4th ACM international workshop on Contents protection and security*. ACM.

30. Winstein, K. 1998. Lexical steganography through adaptive modulation of the word choice hash. Unpublished. http://wwwimsa.edu/~keithw/tlex.

31. Wayner, P. 1992. Mimic functions. *Cryptologia* 16(3): 193–214.

32. Bergmair, R. 2004. Towards linguistic steganography: A systematic investigation of approaches, systems, and issues. Final year thesis, B. Sc.(Hons.) in Computer Studies, The University of Derby.

33. Chapman, M., and G. Davida, 2002. Plausible deniability using automated linguistic steganography. In *Infrastructure Security*, 276–287. Springer.

34. Chapman, M., and G. Davida. 1997. Hiding the hidden: A software system for concealing ciphertext as innocuous text. In *Information and Communications Security*, 335–345.

35. Bergmair, R. 2007. A bibliography of linguistic steganography. In *Proceedings of the SPIE international conference on security, steganography, and watermarking of multimedia contents*. Citeseer.

36. Bender, W., et al. 1996. Techniques for data hiding. IBM Systems Journal, 35(3.4), 313–336.

37. Atallah, M.J., et al. 2001. Natural language processing for information assurance and security: an overview and implementations. In *Proceedings of the 2000 workshop on new security paradigms*. ACM.

38. Nirenburg, S., and V. Raskin, 2004. *Ontological semantics*. MIT Press.

39. Ellegård, A., *English for the computer: The SUSANNE corpus and analytic scheme.*

Chapter 8
Text Watermarking

8.1 Introduction

Any books, article, newspaper, documents, and website are consisted from plain text. Also, plain text is widely used in Internet medium which can exist in all the components of websites, e-books, e-mails, and SMS. Plain text is exposed to many intentional and unintentional attacks, copy, and manipulation. Therefore, protection and security of plain text is a vital requirement. Due to digital watermarking technology being successfully applied for audio, speech, images and video so far, applying digital watermarking technology for text in order to uniquely verify the ownership of the plain text is highly desirable. Text watermarking also protects the content of digital text from illegal copying, copyright violations, redistributions, infringements, and other similar tampering. Although the principles and concepts for text watermarking are similar to image, audio, speech, video watermarking, text watermarking is faced limitation over capacity due to less amount of redundant information than audio, speech, image, and video.

The main requirements for text watermarking are invisibility and uniqueness to provide enough confidentiality and robustness for copyright protection and tamper proofing purposes. In this chapter, the main issues in text watermarking have been discussed. Therefore, the state of the arts in text watermarking are discussed in detail. For this purpose, the text watermarking techniques are explained in taxonomy. Furthermore, different attacks for text watermarking are described.

8.2 Background of Text Watermarking

The plain text has some eminent properties such as language rules, style, semantics, and structure. In addition, some word, line, and block patterns are demarcated between background and foreground of the plain text. Due to binary nature with

© Springer Science+Business Media Singapore 2017
M.A. Nematollahi et al., *Digital Watermarking*, Springer Topics
in Signal Processing 11, DOI 10.1007/978-981-10-2095-7_8

simplest structure of text information, embedding capacity of the plain text is very limited. Some challenges including robustness, imperceptibility, and security are also added to this limitation.

Due to text meaning being valuable, text watermarking should be preserved with the fluency, grammaticality, and meaning of the plain text. Applying semantic transformation can destroy some information in text such as poetry, quotes, author's style, and legality which must be preserved in some situation [1].

8.3 Attacks in Text Watermarking

The lack of robustness and undisclosed technology of the text watermarking has made cyber community enthusiastic to apply text watermarking. For this purpose, analyzing different types of attack for text watermarking is crucial. Text watermarking attacks can be classified into 5 classes including unauthorized deletion, unauthorized detection, unauthorized insertion, reordering, and combination of these attacks. Figure 8.1 presents the well-known attacks which text watermarking should be faced with them.

In unauthorized insertion attack, sentences and words are inserted to the original plain text. In this type of attack, the adversary tries to add another message or watermark in order to insert false information to legal documents. This attack can be protected by embedding time stamps in plain text which incorporate a certifying authority as watermark. In some situation, the adversary tries to hide the

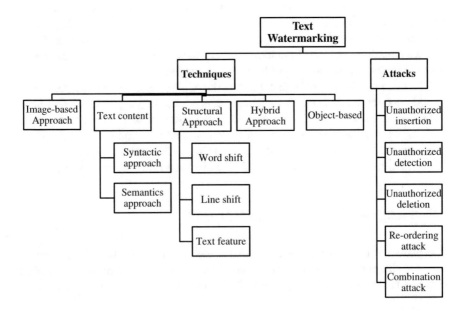

Fig. 8.1 Text watermarking–an overview

author's identity and detract the readers from text originality which is known as unauthorized deletion. For this purpose, some words and sentences are deleted randomly from the original plain text. Protection of text watermarking under unauthorized deletion attack is always required for all the applications. On the other hand, the watermark should be survived during extraction process when attacker altered a number of modifications in text. However, in unauthorized detection attack, the adversary tries to detect whether a watermark is available in plain text. This attack can threaten security for some applications [2].

Sometimes, the words and sentences of the text has been reordered and shuffled in order to destroy the watermark which is known as reordering attack. For instance, the adversary replaces or paraphrases some part of the watermarked text with their synonyms which can affect connotation, writing style, and even text meaning. Finally, these 4 aforementioned attacks can be combined which is known as combination attack. In addition to these 5 attacks, tampering attack is also termed to the combination of reordering, insertion, and deletion attack which has two types such as localized tampering and dispersed tampering. In localized tampering, only a part of the text is modified. However, in dispersed tampering, multiple parts of the text are modified which attempt to make the text look differently. Dispersed tampering is mainly used in literary writings and plagiarism.

8.4 Text Watermarking Robustness

The robustness of text watermarking is computed based on two criteria including watermark distortion rate (WDR) and pattern matching rate (PMR) which are formulized as in Eqs. (8.1) and (8.2):

$$PMR = \frac{\text{Number of patterns correctly matched}}{\text{Number of watermark patterns}} \quad (8.1)$$

$$WDR = 1 - \frac{\text{Number of patterns correctly matched}}{\text{Number of watermark patterns}} \quad (8.2)$$

Both WDR and PMR values are limited to 0 and 1. Whenever the values of PMR are closed to 1 and WDR is closed to 0, the robustness is desirable.

8.5 Text Watermarking Techniques

Text watermarking has firstly emerged in 1994 by Brassil et al. [3, 4]. Recently, due to development of communication and Internet technologies, a number of text watermarking techniques are developed for English, Persian, German, Spanish, and French languages. In this book, text watermarking technique is divided into 5 main approaches such as a syntactic, an image-based, structural, semantic, and hybrid approaches. Figure 8.1 presents these approaches. Each of these techniques is fully described in the following:

8.5.1 Image-Based Approach

As mentioned, text watermarking is a challenge task due to its sensitiveness, low capacity, simplicity, and binary nature. Therefore, in image-based approach, text is treated as image in which watermark is embedded in the appearance and layout of the text image. For embedding watermark in word-shift coding algorithm, the text's words are moved horizontally to expand the space. Moreover, feature coding algorithm is modified by some features of the text image including length of the end lines and pixel of characters. A non-blind text watermarking technique is proposed based on line-shift coding algorithm [3, 5]. This algorithm embedded the watermark bit into the text image by moving lines downward and upward. Although line-shift coding algorithm is robust under diverse attacks, word-shift coding algorithm and feature coding algorithm are easily degraded under diverse attacks.

Another correlation- and centroid-based methods are proposed by treating text as a discrete-time signal [6]. The watermark is embedded by modifying the distance between the centroids of adjacent text blocks or changing the shift direction of adjacent text blocks. An average inter-word algorithm is proposed based on the modification of the distance among the words for each line [7]. This distance can be adjusted based on embedding watermark in phase and frequency of the sine wave. An algorithm based on pixel level is also developed which embed the watermark into the stroke features including serif or width [8]. A text watermarking algorithm is presented based on edge direction histogram manipulation [9].

Inter-word space statistics and word classification approach is developed in [10] which is classifying all of the text's words based on special text features. Then, a segment is comprised based on adjacent words. Each segment is classified based on words' label which is modified by using inter-word statistics to carry the watermark.

An algorithm is developed based on integrated word spaces and inter-character [11]. In addition, the left space and right space of the character is watermarked. For watermark extraction, the value of the hash among character's components after and before watermarking is compared. This algorithm can be extended to word-shifting and line-shifting algorithms.

8.5.2 Syntactic Approach

Basically, characters, words, and sentences are made from the text. Although syntactic approach is mainly used in natural language watermarking, syntactic text watermarking approach has embedded the watermark into the structures of the sentences by applying syntactic transformations [12, 13]. For syntactic text watermarking, firstly, syntactic tree is built then syntactic transformation is used to embed the watermark into the syntactic structure of the transformed text. This technique not only can provide robust and secure text watermarking but also, preserve all the inherent properties of the text which are supported by resources and advances of NLP techniques.

For embedding the watermark, text meaning representation (TMR) is used to manipulate by substitution, pruning, and grafting. Although this approach has good robustness under various attacks, it cannot be applied widely for all the kind of text such as legal documents, contracts, transcripts, and poetry.

Morpho-syntactic alterations text watermarking approach is proposed by embedding the watermark into the functional and the hierarchical dependencies which is transformed by a syntactic tree diagram [14]. Although this approach is proper for agglomerative languages such as, Urdu, Arabic, Korean, and Turkish due to enough space for watermarking, syntactic approach is not adequate for English language. Figure 8.2 presents the watermarking process. Another text watermarking is developed based on adverbial displacement [15]. Syntactical and semantic transformations are applied for watermark embedding [16]. Another effort is done by embedding the watermark into punctuation and ASCII characters of the text [17]. Lastly, three text watermarking are proposed based on NLP techniques which are swapping, lexical substitutions, and shallow parsing. Furthermore, reversible syntactic transformations for data hiding in text have been studied [18].

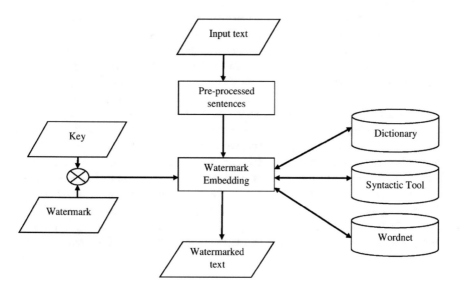

Fig. 8.2 Text watermarking based on syntactic approach

8.5.3 Semantic Approach

Actually, semantic text watermarking is using text semantic structures such as verbs, text contents, grammar rules, words, nouns, and their acronyms, spellings, and sentence structure to embed the watermark into the text. First, semantic approach is developed by synonym substitution technique which replaces specific words with their synonyms [19]. Also, by applying semantic networks, the verbs and nouns of sentence parsed have been changed with certain grammar parser [20].

A recent semantic approach is developed based on common typing errors, abbreviations, and acronyms which are taken place in e-mails, chat, and SMS [21, 22]. For this purpose, linguistic semantic phenomena of presuppositions are applied to observe the discourse representation and meaning. Basically, implicit information in text is considered as presuppositions which are transformed to extraposition, passivization, preposing, and topicalization to carry the watermark. A text watermarking approach is developed for English language by substituting US and UK spellings through the text [23]. Other semantic text watermarking approach is proposed based on light-weighted XoR encryption. In this approach, by removing the static nature of word-abbreviation list, the watermark is embedded into text [24].

8.5.4 Structural Approach

In structural approach, text constituents, their frequency, the ordering, and location are applied to incorporate a watermark instead of modifying the text. For this aim, zero watermarking has been used to utilize characters of original text to construct watermarked text without changing the original text. Basically, language is composed of phrases, clauses, and sentences which are consisted of various types of words including pronoun, noun, verb, adjective, adverb, conjunction, preposition, and acronyms. Even, each word is constructed from alphabet, notation, and symbols. Structural approaches apply syntax and morphology for text watermarking [25]. A structural approach is developed by using double letter [25]. In this approach, firstly, the occurrence of each double letter (AA to ZZ) is counted in each group and the maximum occurring double letter is identified in each group. A MOL (maximum occurring letter) list is formed that contains maximum occurring letter of each group with corresponding group number. Then, partial key is generated from 3-digit group size, 2-digit cipher attributes, and 1-digit cipher choice. The number of sentences to be included in one group is used for 3-digit group size. The watermark extraction is reverse process for embedding and encryption.

8.5.5 *Hybrid Approach*

Hybrid approach is developed to combine different text watermarking approach in order to rectify weakness of each single text watermarking approach. By applying hybrid text watermarking approach, not only the robustness is improved but also watermarking wide text documents are applicable. In this line, structural- and image-based approaches have been combined in a specialized manner [25]. The reason for such combination is due to similarity among pure alphabetical text watermarking.

8.6 Discussion

Nowadays, effective technology like OCR can affect the text watermarking performance. Therefore, image-based text watermarking is a failure approach under text reproduction and retyping attacks. Also, copy–paste to notepad can also destroy this approach.

On the other hand, combination of NLP with syntactic approach can give an efficient text watermarking approach. However, the research in NLP is not progressing rapidly which cannot provide enough support for syntactic approach. Moreover, transformations in NLP are usually irreversible which degraded the syntactic analyzers. Although semantic approach can improve capacity of watermarking and increase impressiveness of watermarking, semantic approach is considered as impractical and conceptual approach. In addition, applying semantic approach for sensitive text including poetry, quotation, and legal documents is not possible due to value and semantic connotation of the text. The robustness of synonym-based approach is also vulnerable when synonym is substituted randomly. Furthermore, it is requiring huge collection databases for a large synonym dictionary. Apart from these limitations, not only NLP is immature for semantic analyzers but also semantic approach is language-dependent which is varying with time and dynamic nature of the language. On the other hand, structural approach can only be used for alphabetical and image watermark which is not proper for all the different types of text documents. Furthermore, other essential text constituents such as conjunctions, prepositions, verbs, and nouns are not applied for structure text watermarking which can reduce the capacity. However, hybrid approach can successfully compose text watermarking approaches for enhancing text security, imperceptibility, capacity and provide better robustness.

8.7 Summary

Text is important medium of information exchange. Therefore, text requires to be completely protected against copyright violators like deletion, reordering, and insertion attacks. This protection can be provided by practical and efficient text watermarking approaches. This chapter has exclusively described text watermarking in detail. The challenges, the application areas, and possible attacks are also discussed. It is inferred that lack of integrity, robustness, generality, and accuracy for text watermarking are caused less research progress for developing text watermarking approaches. Therefore, a huge gap is available for research in this area. It can be concluded that neither image-based approach is robust against reproduction attacks nor syntactic and semantic approaches can provide robustness under random tampering attacks. For future work in text watermarking approaches, the rest of the challenges such as language specific, applicability, computationally expensive, and usability are remained to be solved.

References

1. Topkara, M., et al. 2006. Natural language watermarking: Challenges in building a practical system. In *Electronic imaging 2006*. International Society for Optics and Photonics.
2. Zhou, X., et al. 2009. Security theory and attack analysis for text watermarking. In *E-Business and information system security. International conference on EBISS'09* 2009. IEEE.
3. Brassil, J.T., et al. 1995. Electronic marking and identification techniques to discourage document copying. *IEEE Journal on Selected Areas in Communications,* 13(8): 1495–1504.
4. Brassil, J., et al. 1995. Hiding information in document images. In *Proceedings of conference on information sciences and systems (CISS-95)*. 1995. Citeseer.
5. Khan, A., and A.M. Mirza. 2007. Genetic perceptual shaping: Utilizing cover image and conceivable attack information during watermark embedding. *Information fusion* 8(4): 354–365.
6. Low, S.H., N.F. Maxemchuk, and A.M. Lapone. 1998. Document identification for copyright protection using centroid detection. *IEEE Transactions on Communications* 46(3): 372–383.
7. Low, S.H., and N.F. Maxemchuk. 2000. Capacity of text marking channel. *IEEE Signal Processing Letters* 7(12): 345–347.
8. Amano, T., and D. Misaki. 1999. *A feature calibration method for watermarking of document images*. In *Document analysis and recognition. Proceedings of the fifth international conference on ICDAR'99*. IEEE.
9. Kim, Y.-W., and I.-S. Oh. 2004. Watermarking text document images using edge direction histograms. *Pattern Recognition Letters* 25(11): 1243–1251.
10. Kim, Y.-W., K.-A. Moon, and I.-S. Oh. 2003. *A text watermarking algorithm based on word classification and inter-word space statistics*. In *Null* 2003. IEEE.
11. Yang, H., and A.C. Kot. 2004. Text document authentication by integrating inter character and word spaces watermarking. In *Multimedia and expo, 2004. IEEE international conference on ICME'04*. IEEE.
12. Atallah, M.J., et al. 20014. Natural language watermarking: Design, analysis, and a proof-of-concept implementation. In *Information hiding*. Berlin. Springer.
13. Atallah, M.J., et al. 2001. Natural language processing for information assurance and security: An overview and implementations. In *Proceedings of the 2000 workshop on new security paradigms*. ACM.

14. Meral, H.M., et al. 2009. Natural language watermarking via morphosyntactic alterations. *Computer Speech & Language* 23(1): 107–125.
15. Kim, M.-Y. 2010. Natural language watermarking for Korean using adverbial displacement. In *Multimedia and ubiquitous engineering. International conference on MUE 2008*. IEEE.
16. Hoehn, H. 2007. Natural language watermarking. In *Seminar series selected topics of IT security, summer term*.
17. Kankanhalli, M.S., and K. Hau. 2002. Watermarking of electronic text documents. *Electronic Commerce Research* 2(1–2): 169–187.
18. Murphy, B., and C. Vogel. 2007. Statistically-constrained shallow text marking: techniques, evaluation paradigm and results. In *Electronic imaging 2007*. International Society for Optics and Photonics.
19. Topkara, U., M. Topkara, and M.J. Atallah. 2006. The hiding virtues of ambiguity: quantifiably resilient watermarking of natural language text through synonym substitutions. In *Proceedings of the 8th workshop on multimedia and security*. ACM.
20. Sun, X., and A.J. Asiimwe. 2005. Noun-verb based technique of text watermarking using recursive decent semantic net parsers, In *Advances in natural computation*, 968–971. Berlin: Springer.
21. Topkara, M.K. 2007. *New designs for improving the efficiency and resilience of natural language watermarking*: ProQuest.
22. Vybornova, O., and B. Macq. 2007. A method of text watermarking using presuppositions. In *Electronic imaging*. International Society for Optics and Photonics.
23. Shirali-Shahreza, M. 2008. Text steganography by changing words spelling. In *Advanced communication technology. 10th international conference on ICACT 2008*. IEEE.
24. Rafat, K.F. 2009. Enhanced text steganography in SMS. In *Computer, control and communication. 2nd international conference on IC4*. IEEE.
25. Jalil, Z. 2010. *Copyright protection of plain text using digital watermarking*. National University of Computer and Emerging Sciences, Islamabad, Pakistan.

Chapter 9
Software Watermarking

9.1 Introduction

Generally, software watermarking is considered as a branch of digital watermarking technique which has not been attracted enough attention by researcher. Since digital watermarking is began in 1954 [1], the major effort on software watermarking is started from 1990 [2]. Although preliminary concepts in software watermarking have been proposed in some literatures [3–5], the first main software watermarking algorithm is a patented one [6].

Software watermarking is the art and science of embedding piece of secret code into software program to proof the ownership without changing the semantic operation of the program code. The proof of ownership can be claimed by extracting the secret information from the software program when unauthorized version of the software has been used. Software watermarking has some application in fingerprint mark, licensing mark, validation mark, and authorship mark [7].

In this chapter, the state of the arts in software watermarking is discussed in detail. For this purpose, the software watermarking techniques are explained in a taxonomy. Furthermore, different attacks for software watermarking are described.

9.2 Background of Software Watermarking

Software watermarking is relatively a new trend in software engineering which tries to secure the computer programing code from any misuse. In contrast to multimedia watermarking which does not have operating semantics layer, software watermarking must not have any modification on operational and task of the computer program. Similar to multimedia watermarking, it is possible to embed the watermark into the appearance of software object which is known as

© Springer Science+Business Media Singapore 2017
M.A. Nematollahi et al., *Digital Watermarking*, Springer Topics
in Signal Processing 11, DOI 10.1007/978-981-10-2095-7_9

"Easter Eggs". However, there are other possibilities for embedding the watermark. Figure 9.1 shows a simple example for software watermarking by inserting a string into the code.

Software watermarking can be classified into different ways such as functionalities, properties, extractions, robustness, applications, and visibility. Figure 9.2 shows various classification of software watermarking. As shown, software watermarking trends are overviewed as five classes including application, extraction method, robustness, visibility, and blindness. From application perspective, software watermarking can be viewed as four main groups:

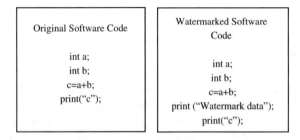

Fig. 9.1 An example of a software watermarking

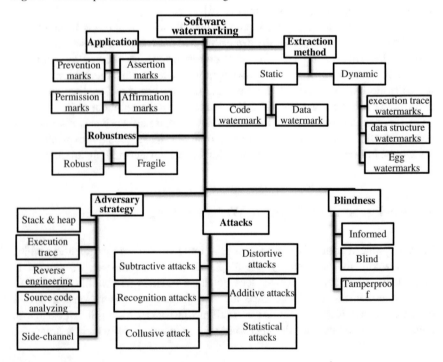

Fig. 9.2 Classification in software watermarking technology

(a). Prevention marks which prevents unauthorized access to software.
(b). Assertion marks which construct a claim on software ownership publicly.
(c). Permission marks which only allows limited access or modification on operational copy of the software.
(d). Affirmation marks which provide authenticity for each end user.

From the extraction method's perspective, software watermarking is divided to two main classes such as static and dynamic. In static approach, the software does not need to run due to the watermark is embedded into the data or code which are known as data watermark and code watermark, respectively. However, dynamic watermarking inserts some watermark codes in the objects of the software. While the software is running, the watermark is extracted. There are three approaches for dynamic watermarking such as data structure, execution trace, and Easter Egg. From robustness perspective, software watermarking is divided to three categories such as robust, tamperproof, and fragile. The robust software watermarking is applied for presentation and assertion applications due to the watermark has to be extracted when adversary attacks such as near semantics-preserving code translation and casual semantics preserving are taken place maliciously. In situations that a skilled adversary is going to tamper the software, tamperproof software watermarking can be applied as tamper detection method. However, fragile software watermark is applied for permission and affirmations applications due to its characteristic which is destroyed when software is modified.

From blindness perspective, software watermarking is divided into two methods including informed and blind which are related to the availability of the original software and watermarked software in the software extraction side. If both of them are required during watermark extraction, it is known as informed software watermarking technique. Otherwise, it is known as blind software watermarking technique when only watermarking software is available.

Currently, software watermarking protection techniques have been used for different types of attacks. Obfuscation and tamper-proofing are the two main mechanisms to protect the software watermark. Obfuscation is a technique to translate a software code into another software code, and somehow semantics of the software code is preserved. This improves the security of the software watermarked code from adversary attacks due to the analysis of the obfuscation software code is difficult. However, in tamper-proofing, it is difficult for adversary to remove the tamper-proof software program without affecting the software's usability.

There are four main platforms for software watermarking. Table 9.1 summarizes each platform.

9.3 Formal Representation of Software Watermarking

Software watermarking is discussed in a formal way in terms of embedding and extraction which are known as encoding function and exposition function [12].

Table 9.1 The main platforms for software watermarking

Research platforms	Details
JavaWiz [8]	It is developed in Purdue University based on Java and Curve Tracing (CT) algorithm which can watermarked the java source codes
Hydan [9]	It is developed in Columbia University for watermarking executable codes
UWStego [10]	It is developed in University of Wisconsin for designing, testing and developing new software watermarking through its toolset
SandMark [11]	It is developed in University of Arizona based on java to provide comprehensive research tools

Definition 9.1 A watermark stream of 0 and 1 bit with finite length which can be defined as follows: $W = \{0, 1\}^N, N \geq 0$.

Definition 9.2 Let P denotes software program code, W denotes watermark, $\hat{P}$ denotes watermarked software program code, and an embedding function is defined as follows: $\mathrm{EMB}(P, W) \rightarrow \hat{P}, P \in \hat{P}$.

Definition 9.3 Let P denotes original software program code, $\hat{P}$ denotes watermarked software program code, W denotes embedded watermark, $\hat{W}$ denotes extracted watermark and an extraction function is defined as follows: $\mathrm{EXT}\left(P, \hat{P}\right) \rightarrow \hat{W}, W = \hat{W}$. If P does not require $\mathrm{EXT}(.,.)$, the software watermark is blind. Otherwise, it is informed software watermark.

Definition 9.4 If two temporary variables interfere in the software code, there is an edge between them in the interference graph.

Definition 9.5 Let $\mathrm{dom}(P)$ be the set of sequence accepted by software program P, $\mathrm{Out}(P, I)$ be the output of software program P for input I, $S(P, I)$ be state of software program P for input I, $|S(P, I)|$ be the size of state S, and T be the software transformations (including T_{sem} semantics-preserving transformation, T_{stat} state-preserving transformation, and T_{crop} cropping transformation). T_{sem} can preserve the behavior of the watermarked software program, T_{stat} can preserve the internal state of the watermarked software program (like code optimization transformation $T_{\mathrm{stat}} \subset T_{\mathrm{sem}}$), and T_{crop} cannot preserve both behavior and state of the watermarked software program. In the following, each of these transformations is formally expressed as in Eqs. (9.1)–(9.3).

$$T_{\mathrm{sem}} = \{t : T | P \in p, I\varepsilon\mathrm{dom}(P), \mathrm{dom}(P) \in \mathrm{dom}(t(P)), \mathrm{out}(P, I) = \mathrm{out}(t(P), I)\} \tag{9.1}$$

$$T_{\mathrm{stat}} = \{t : T | P \in p, I \in \mathrm{dom}(P), S(P, I) = S(t(P), I)\} \tag{9.2}$$

$$T_{crop} = \{t : T | \exists P \in p, \exists I \in \mathrm{dom}(P), I \in \mathrm{dom}(P), (I \notin \mathrm{dom}(t(P)) \vee \mathrm{out}(P, I) \neq \mathrm{out}(t(P), I))\} \tag{9.3}$$

The recognizer R_T can be expressed as in Eq. (9.4)

$$R_T : (P \times S) \rightarrow W$$

$$\forall t \in T : p\left(R_T\left(t\left(\hat{P}\right), S\left(t\left(\hat{P}\right), I\right)\right)\right) = p(w) \tag{9.4}$$

Therefore, different set of recognizer R_T including trivial, strong, ideal, static, and pure recognizers can be defined in the following:

Trivial recognizer (fragile software recognizer): The W cannot be detected if any transformation is taking place on $\hat{P}$ which is expressed as in Eq. (9.5)

$$R_\phi(\hat{P}, S(\hat{P}, I)) \tag{9.5}$$

Ideal recognizer (robust software recognizer): It is resilient against any types of transformation which is expressed as in Eq. (9.6)

$$R_T(\hat{P}, S(\hat{P}, I)) \tag{9.6}$$

Strong recognizer: It is robust in terms of semantics-preserving transformation T_{sem} which is expressed as in Eq. (9.7)

$$R_{T_{\text{sem}}}(\hat{P}, S(\hat{P}, I)) \tag{9.7}$$

Pure dynamic: It is not examining the text of $\hat{P}$ and it is only examine the execute state of $\hat{P}$ which is expressed as in Eq. (9.8)

$$R_T(\phi, S(\hat{P}, I)) \tag{9.8}$$

9.4 Software Watermarking Criteria

Similar to multimedia watermarking criterion, software watermarking techniques are also evaluated in terms of some criteria resilience (robustness), data rate (capacity or payload), stealth (imperceptibility), and performance criterion. Figure 9.3 shows the important criterion for evaluating and validating the performance of the software watermarking technique. Figure 9.3 shows the main criterion in software watermarking.

As shown from Fig. 9.3, four criteria such as data rate which is referred to ratio between watermark and watermarked software program; resilience is referred to robustness of the software watermarking under semantics-preserving translations; stealth is referred to invisibility over static features of original software program and watermarked software program; and performance criterion is referred to ratio between size and execution time of the software watermarked program to original software program. In the following, formal definition of each criteria are expressed.

Fig. 9.3 Different criterion in software watermarking

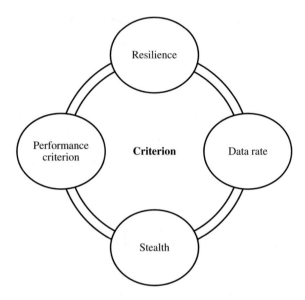

Definition 9.6 A watermark W in watermarked software program $\hat{P}$ is statically stealth if for a statistical measure M, $M(P) - M\left(\hat{P}\right) < \Omega$ and it is dynamical stealth if $M(S(P, I)) - M\left(S(\hat{P}, I)\right) < \Omega$. Ω corresponds to a very small value.

Definition 9.7 Let $|P|$ be the size of software program in words, $H(w) = \log_2^{|W|}$ be the entropy of the watermark W, and $|S(P)| = \mathrm{Max}_{I \in \mathrm{dom}(P)}|S(P, I)|$ which is the least upper bound of internal state S for program P and input I. The static data rate as in Eq. (9.9):

$$\frac{H(w)}{\max(1, \left|\hat{P}\right| - |P|)} \geq 1 \tag{9.9}$$

The dynamic data rate is expressed as in Eq. (9.10):

$$\frac{H(w)}{\max(1, \left|S(\hat{P})\right| - |S(P)|)} \geq 1 \tag{9.10}$$

Definition 9.8 The watermarked software program $\hat{P}$ is space and size resiliencies if the recognizer R_T following expression [as in Eq. (9.11)] is satisfied.

$$\forall t \in T : (p(R_T\left(t\left(\hat{P}\right), S\left(t\left(\hat{P}\right), I\right)\right)) \neq p(w) \Rightarrow \left\{ \begin{array}{ll} \dfrac{\left|S(t\left(\hat{P}\right),I)\right|}{\left|S(\hat{P},I)\right|} \geq r & \text{Space resilience} \\[4mm] \dfrac{\left|t(\hat{P})\right|}{\left|\hat{P}\right|} \geq r & \text{Size resilience} \end{array} \right. \tag{9.11}$$

where r measures the weakness of the watermarked.

Definition 9.9 The watermarked software program $\hat{P}$ is runtime resilient (performance), if the recognizer R_T following expression [as in Eq. (9.12)] is satisfied.

$$\forall t \in T : (p(R_T\left(t\left(\hat{P}\right), S\left(\hat{P}, I\right)\right)) \neq p(w)\exists i \in \text{dom}(P)\frac{\text{Time}(t\left(\hat{P}\right), i)}{\text{Time}(\hat{P}, i)} \geq r \quad (9.12)$$

9.5 Software Watermarking Techniques

In this section, the major software watermarking techniques are described. Figure 9.4 classified different watermarking techniques based on their methodology.

In the following, each software watermarking technique shown in Fig. 9.4 is explained in detail.

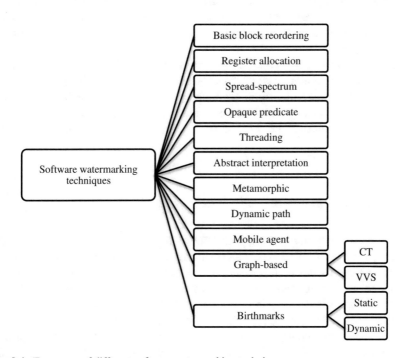

Fig. 9.4 Taxonomy of different software watermarking techniques

9.5.1 Basic Blocking Reordering Technique

In this technique, a basic block in software program, where a set of sequential instructions with a single entry and a single exit points, are reordering to embed the watermark [6]. For this purpose, firstly, some executable basic blocks are selected as a group. Then, a watermark is embedded by reordering them without affecting on execution and function of the original software program. For extraction the watermark, the orders of these basic blocks are checked to detect the watermark. However, the resilience of this technique is low against software attacks like watermarking the software again. The resilience of this technique may be improved by applying opaque predicate to make false dependencies among basic blocks which is difficult to eliminate them.

9.5.2 Register Allocation Technique

In this technique, software watermarking is considered as a constraint problem or graph coloring problem. Basically, graph coloring problem is a special case of graph labelling problems which try to select minimum colors for coloring the vertices of the graph in such a way that there is no two adjacent vertices with the same color [13, 14].

Actually register allocation technique is applying Quadratic Programming (QP) for software watermarking [15]. However, a research has been conducted to show that not only QP algorithm cannot provide good resilience due to the flaws in its nature but also, second attempt through Quantitative Perfusion SPECT (QPS) cannot improve the resilience of the software watermarking against attacks [16].

9.5.3 Spread Spectrum Technique

Spread spectrum watermarking technique is mainly developed for multimedia objects. However, in software watermarking, based on speaker spectrum three main steps are included: representation detection, watermark embedding, and watermark testing. A spread spectrum approach has been presented, which considered software code as a statistical object [17]. A vector extraction paradigm (VEP) is used to extract the representation group from the software code. Then, speaker spectrum technique embeds the watermark inside it. Another approach is proposed based on changing the graph depth of the software code [18]. Then, the watermark is extracted when certain particular input is given to the software.

9.5.4 Opaque Predicate Technique

Opaque predicate is referring to prediction and expectation of outcome of the software by the programmer at prior, but it needs to evaluate at runtime too. Opaque predicate technique can also be used for software watermarking and software obfuscation. An opaque predicate approach is proposed by defining three steps. Firstly, java code is appended by a dummy method. Next, opaque predicate has applied to add this dummy invocation to source software code. Finally, watermark is embedded based on stealthy technique by modifying bit sequences of dummy which replaces opcodes and overwrites numerical operands [19].

9.5.5 Threading Technique

Threading technique mainly is applied for multi-threading software program based on the advantage of NT algorithm which is intrinsic randomness of a thread. NT algorithm embeds the watermark inside the software program by applying two steps. First, suitable locks are found by increasing the number of possible path executions in software program. This step preserves the semantics of the software program. Second, ensure that only small subsets of these added locks are executed in watermarked software program. This technique can provide enough resilience due to the difficulty in analyzing multi-thread software program code. An approach for Java byte code is proposed based on threading technique [20].

9.5.6 Abstract Interpretation Technique

Abstract interpretation technique is embedding the watermark in software program by assigning the value of the designate integer local variables. During execution, the watermark is extracted by analyzing these values which is done by abstract interpretation framework even under small portion of the watermarked software program [21].

9.5.7 Metamorphic Technique

Metamorphic technique is increasing the diversity of program code by applying metamorphic code transformation in order to embed a pattern of the watermark into the software program. However, this technique cannot provide resilience under attacks [22].

9.5.8 Dynamic Path Technique

Dynamic path technique embedded the watermark into the runtime branch structure of the software program by using two main steps. First, suitable points in
software program are determined by tracing its dynamic behavior. This tracing
can be done by applying an input sequence for finding executing path of the software program. Second, a secret input sequence as the watermark is embedded
into the suitable points of the branch sequences. For watermark extraction, the
secret input sequence is used to trace and check the branch sequence to detect the
watermark.

Due to branch structure captures huge semantics of the software program and
it is necessary part of the software program, analyzing, and observation of the
branch structure completely is very difficult. Therefore, this technique can provide
enough resilience under attacks [23].

9.5.9 Mobile Agent Technique

In this technique, mobile agent is watermarked which assure the integrity of the
mobile agent during its execution in all host sides. Therefore, the result of each
agent execution is transferred to original host to verify the integrity of the mobile
agent execution overall [24].

9.5.10 Graph-Based Technique

In this technique, the watermark is mapped into a special kind of graph which
can be static (VVS algorithm) or dynamic (CT algorithm). In VVT software
watermarking technique, the control flow graph of the original software program is modified for embedding the watermark [25]. However, in CT software
watermarking, during execution the data structure graph is watermarked based
on three main steps. First step is finding the best graph for embedding the watermark. Whenever the graph is selected properly, the stealthy of the software
watermarking is increased. Second step is dividing the candidate graph to several
sub-graphs. Third step is embedding the watermark to generated code of sub-
graph by using CT algorithm [26, 27]. Generally, CT algorithm not only provide
the stealthily but also can correct the error during watermark extraction when a
suitable graph is selected. However, the adversary can use CT algorithm disadvantages, which independency between original software code and graph generated code, to remove the watermark without any serious degradation on software
behavior.

9.5.11 Birthmarks Technique

In contrast to software watermark that embedded certain features into the software code purposely, software birthmark is inherited from its own characteristic of the software code. Basically, both software watermark and software birthmarks can be combined in order to improve the protection of the software code. Software theft for Java classes can be detected by concept of Java birthmark which is based on unique characteristics set of Java class [28]. For example, if same birthmarks are found for two Java classes, both classes have the same source which means one of them is copied. There are two types of software birthmarks such as static and dynamic. Four static software birthmarks are developed in literature includes the sequence of method calls, used classes, the inheritance structure, and constant values in field. However, sequence API function, frequency of API function calls, and the full path or control flow of the software code are the main dynamic birthmarks. It must be mentioned that credibility and resilient are important for software birthmark. The former ensures that the software do not generate false positives and the latter ensures the software is not seriously transformed through semantic transformations.

9.6 Attacks in Software Watermarking

Many attacks can threat software code at client and host sides. Software watermarking can only protect the software against attacks at host side. As seen in Fig. 9.2, many attacks such as recognition, distortive, subtractive, statistics, collusive, and additive attacks are available which should be considered during software watermark development phase. These attacks are mainly implemented by five adversary strategies including stack and heap, execution trace analysis, source code analysis, side-channel, and reverse engineering [29]. In stack and heap approach, the amount of consumed memory for both stack and heap of the software watermarked code are analyzing for detecting the watermark. Similarly, execution trace analysis approach tries to detect the watermark by analyzing the watermarked execution history (such as decisions, branch points, and function entries). However, source code approach tries to discovering the watermark by similar way for find bugs in software. In the side-channel, the location of the watermark is revealed by using side-channel techniques (e.g., using power consumption signal, or leakage generators, or clock cycle). Reverse engineering is a basic approach to analyze functionality and features of the watermarked software code when original software code is not available. In the following, each of these attacks is described.

Additive attack: In this attack, an attacker tries to insert a new watermark to the watermarked software to doubt about ownership. On the other hand, the real ownership is under the question by an adversary.

Subtractive (cropping) attack: In this attack, an attacker tries to discover and remove the watermark from the watermarked software code by preserving the main software functionality. This attack is the final aim of any adversary and is in opposite of additive attack.

Distortive attack: In this attack, an attacker tries to distort watermark extraction process by modifying the watermarked software code. On the other words, the main functionality of the watermarked software is preserved but the ownership is under the question. These attacks are mainly implemented by obfuscation, optimization, and translation transformation which preserves semantics of the software code.

Recognition attacks: In this attack, an attacker tries to recognize the watermark extraction process for misleading watermark detection result. For this purpose, the attacker modifies some parts in watermark extraction process to convince the court that his/her watermark detector is the real one for detecting the ownership.

Statistical attack: In this attack, an attacker tries to identify anomalies among commands, codes, and computations in order to locate the watermark.

Collusive attack: In this attack, different copies of the watermarked software code are compared for identifying the location of the watermark.

9.7 Discussion

Available software watermarking techniques have not been a developed robust technique that can withstand against all types of attacks [30]. Therefore, many basic and fundamental issues are remained to be solved for proposing software watermark technique. Table 9.2 provides a comparison among different software watermarking techniques (Some parts are Table 9.2 are got from [31]). As seen, none of the software watermarking technique is immune against all attacks. Static

Table 9.2 Comparison among various software watermarking techniques

Approach	Technique	Resiliency	Overhead	Part protection	Stealth	Data rate	Credibility
Static	Data watermark	L	L	L	L	H	H
	Code watermark	M	M	H	H	M	M
Dynamic	Easter egg	H	M	M	L	M	M
	Data structure	H	L	H	M	H	H
	Execution trace	M	H	H	M	M	L

H = high
M = Medium
L = Low

software watermarking techniques cannot provide enough resilient against distortive attacks, but it is easy to use them through duplication for different parts of the software program. However, dynamic software watermarking techniques can provide proper stealth especially for object-oriented software which requires high-complex heap structures. Although dynamic graph and Easter Egg techniques are resilient in terms of various distortive attacks, it seems they cannot protect software code against illegal reuse of the valuable component and subtractive attacks. Apart from these attacks, obfuscating transformation techniques can guarantee the resilient probably all of the software watermarking techniques against collusive attacks due to fluidity of the software program. On the other words, the instructions in a software program can easily sweeping without changing the software behavior, functionality, and semantics which are quite difficult for multimedia watermarking.

References

1. Cox, I.J., et al. 1996. A secure, robust watermark for multimedia. In *Information hiding*. Springer.
2. Tamada, H., et al. 2004. Design and evaluation of birthmarks for detecting theft of java programs. In *IASTED conference on software engineering*.
3. Samson, P.R. 1994. Apparatus and method for serializing and validating copies of computer software. Google Patents.
4. Moskowitz, S.A. and M. Cooperman. 1998. Method for stega-cipher protection of computer code. Google Patents.
5. Grover, D. 1992. *Protection of computer software: Its technology and application.* Cambridge University Press.
6. Davidson, R.I. and N. Myhrvold. 1996. Method and system for generating and auditing a signature for a computer program. Google Patents.
7. Bhat, V., I. Sengupta, and A. Das. 2010. An adaptive audio watermarking based on the singular value decomposition in the wavelet domain. *Digital Signal Processing* 20(6): 1547–1558.
8. Palsberg, J., et al. 2000. Experience with software watermarking. In *16th Annual conference on computer security applications, 2000, ACSAC'00.* IEEE.
9. El-Khalil, R. and A. Keromytis. 2004. *Hydan: Embedding secrets in program binaries.*
10. Collberg, C., et al. 2004. *Uwstego: A general architecture for software watermarking.*
11. Collberg, C., G. Myles, and A. Huntwork. 2003. Sandmark–A tool for software protection research. *IEEE Security and Privacy* 4: 40–49.
12. Ahmed, M.A., et al. 2010. A novel embedding method to increase capacity and robustness of low-bit encoding audio steganography technique using noise gate software logic algorithm. *Journal of Applied Science* 10(1): 59–64.
13. Wong, J.L., G. Qu, and M. Potkonjak. 2004. Optimization-intensive watermarking techniques for decision problems. *Computer-Aided Design of Integrated Circuits and Systems, IEEE Transactions on* 23(1): 119–127.
14. Qu, G. and M. Potkonjak. 2000. Hiding signatures in graph coloring solutions. In *Information hiding*. Springer.
15. Myles, G. and C. Collberg. 2004. Software watermarking through register allocation: Implementation, analysis, and attacks. In *Information security and cryptology-ICISC 2003,* 274–293. Springer.

16. Zhu, W. and C. Thomborson. 2006. Algorithms to watermark software through register allo-
    cation. In *Digital rights management. Technologies, issues, challenges and systems,* 180–191.
    Springer.
17. Stern, J.P., et al. 2000. Robust object watermarking: Application to code. In *Information hid-
    ing.* Springer.
18. Curran, D., et al. 2004. Dependency in software watermarking. In *Proceedings of interna-
    tional conference on information and communication technologies: from theory to applica-
    tions,* 2004. IEEE.
19. Monden, A., et al. 2000. A practical method for watermarking java programs. In *COMPSAC
    2000. The 24th annual international on computer software and applications conference,
    2000.* IEEE.
20. Al-Haj, A. 2014. An imperceptible and robust audio watermarking algorithm. *EURASIP
    Journal on Audio, Speech, and Music Processing* 2014(1): 1–12.
21. Kang, G.S., T.M. Moran, and D.A. Heide. 2005. *Hiding information under speech.* DTIC
    Document.
22. Cheung, W. 2000. Digital image watermarking in spatial and transform domains. In
    Proceedings of TENCON 2000. IEEE.
23. Kim, H.J., and Y.H. Choi. 2003. A novel echo-hiding scheme with backward and forward
    kernels. *IEEE Transactions on Circuits and Systems for Video Technology* 13(8): 885–889.
24. Guerchi, D., et al. 2008. Speech secrecy: An FFT-based approach. *International Journal of
    Mathematics and Computer Science* 3(2): 1–19.
25. Ma, B., et al. 2014. Secure multimodal biometric authentication with wavelet quantization
    based fingerprint watermarking. *Multimedia tools and applications* 72(1): 637–666.
26. Cvejic, N. and T. Seppanen. 2004. Reduced distortion bit-modification for LSB audio steg-
    anography. In *2004 7th International conference on signal processing, 2004. Proceedings of
    ICSP'04.* IEEE.
27. Matsuoka, H. 2006. Spread spectrum audio steganography using sub-band phase shifting. In
    *International conference on intelligent information hiding and multimedia signal processing,
    2006. IIH-MSP'06.* IEEE.
28. Zong, T., et al. 2014. *Robust histogram shape based method for image watermarking.*
29. Zhu, W.F. 2007. *Concepts and techniques in software watermarking and obfuscation.*
    Auckland: ResearchSpace.
30. Georg T. Becker, D.S., Christof Paar and Wayne Burleson. 2012. Detecting software theft in
    embedded systems: A side-channel approach. *IEEE Transactions on Information Forensics
    and Security* 7(4): 1144–1154.
31. Lim, H.-I. 2015. A performance comparison on characteristics of static and dynamic software
    watermarking methods. *Indian Journal of Science and Technology* 8(21).

Chapter 10
Relational Database, XML, and Ontology Watermarking

10.1 Introduction

Today, digital watermarking technology has emerged as an effective tool for relational databases and eXtensible Mark-up Language (XML) data in order to protect the copyright, detect tamper, trace traitor, and maintain the integrity of the data. The conventional multimedia watermarking techniques may not be suitable enough to apply for relational databases and XML data due to some differences between them which are described in the following:

(i)   Small capacity for watermarking due to less redundant data: In contrast to multimedia objects that contain large amount of bits for embedding the watermark, the amount of bits in database's tuples, which are defined as a collection of independent object, are very limited for embedding the watermark. Therefore, the watermark capacity for database watermarking is most difficult than multimedia watermarking.

(ii) The nature of relational data: Generally, the relative temporal and the spatial domains for multimedia objects may not be seriously changed which is consistent. However, the position of data in relational database is out of order due to the lack of order for each set, which is defined as a collection of tuples.

(iii) Frequent database operations: In contrast to multimedia content that is not updating, replacing, and dropping in the normal operations, the tuples in relational database are updating, inserting, and dropping which are very normal operations for database systems.

(iv) Small imperceptibility for watermarking due to the lack of any psychophysical models: In contrast to multimedia watermarking techniques that apply HAS and HVS models to embed the watermark, there is not such a model for relational database to be exploited for watermarking.

© Springer Science+Business Media Singapore 2017

M.A. Nematollahi et al., *Digital Watermarking*, Springer Topics in Signal Processing 11, DOI 10.1007/978-981-10-2095-7_10

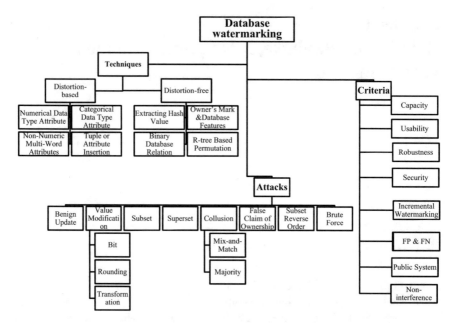

Fig. 10.1 An overview of database watermarking

Figure 10.1 illustrates an overview on different issues in relational database watermarking.

10.2 Issues in Relational Database Watermarking

Similar to criterion for multimedia watermarking including capacity, robustness, FPR, FNR, and imperceptibility, there are other issues for relational database watermarking which should be considered. In the following, some of these issues are explained:

- Usability: Embedding the watermark into the tuples and sets of the relational database should not change data of the database somehow operation and usability of the relational database system is degraded. Depending on the nature of the stored records, the amount of usability is tolerating from one database to another. There are some constraint functions including structural constrains, semantics constrains, and measurement constraints which can check watermarked database usability. The basic and common criterion for statistical measurement constraints are standard deviation and mean of the data. However, input SQL statements which are defined by user based on relational table can be applied for structural constraints and semantics constraints [1].
- Security: The overall database security is relied on some secret parameters which must be protected by database administrator (DBA) or database owner.

- Blindness: Ability of detecting the watermark from irrespective of later updates to the original relation and a copy of the database relation (which known as blindness) is a critical factor for database watermarking.
- Incremental Watermarking: After watermarking relational database, there is a need to unaltered the untouched watermarked tuples and watermarking the tuples that are only modified and added.
- Non-interference: It should be possibility to embed multiple watermarks into a single relational database without any interference between them.
- Public System: The watermarking relational database is performed publicly. Thus, it should be constructed based on private parameters which are only defense mechanism [1].

10.3 Database Watermarking Attacks

Basically, many intentional and unintentional attacks are available that erase and damage the robustness and security of watermark in relational database. Therefore, there is a requirement to known these attacks properly which are described in the following:

I.   Benign Update: There are some unintentional operations including update, delete, and add for relational database which may degrade the extraction process of the watermark. For example, some of the watermark bits are undetectable during the update tuples of the database which causes erroneously flipped the data.

II. Value Modification Attack: This attack can be divided into three main category including bit attack, rounding attack, and transformation attack.

   a. Bit attack: In this attack, some of the watermarked data bits are altered in order to destroy the watermark. However, this alternation cannot be huge due to make the whole data completely usefulness. Whenever, the positions of the watermark bits are known, this type of attack can be more destructive. This attack can be implemented by changing certain bit positions randomly (called randomization attack), setting certain bit positions to zero (called zero out attack), or inverting certain bit positions (called bit flipping attack).

   b. Rounding attack: This attack is rounding numeric attribute's values to certain rate in order to remove watermark from numerical values in database. Whatever this rate can estimate the bit positions of the watermark correctly, this attack would be successful. Overestimation can destroy the usefulness of the data while underestimation makes the attack completely unsuccessful.

   c. Transformation attack: In contrast to rounding attack that linearly transformed the numerical values, in transformation attack, a unit of measurement is converted to another unit (e.g., centimeter to inch).

III.   Subset Attack: This attack is based on updating and deleting a subset of the attributes and tuples of the watermarked.

IV.   Superset Attack: This attack is adding some new records and tuples to the watermarked database in order to disturb the watermarking extraction process.

V.   Collusion Attack: This attack is taking place when multiple copies of the database are available for adversary. Two main collusion attacks are including:

    a. Mix-and-Match Attack: In this attack, an adversary attempts to create a fake relation by applying multiple relations with similar information for disjointing tuples and records.

    b. Majority Attack: In this attack, an adversary creates a fake relation based on majority function which computes every corresponding bit value of the fake relation from all copies of the databases in such a way that even owner cannot detect it.

VI.   False Claim of Ownership: This attack is increasing doubts about ownership of the data by given an attacker with evidence.

    a. Additive Attack: In this attack, an attacker inserts a new watermark in watermarked database in order to overwrite a new ownership.

    b. Invertibility Attack: This attack claims a fake ownership by discovering a random occurrence from the watermarked database.

VII.   Subset Reverse Order Attack: This attack reorders the positions of the tuples and records for a relation for disturbing the watermark extraction process.

VIII.   Brute Force Attack: This attack attempts to search all values for private parameters in a possible space. Whenever this space is big, the possibility of finding correct values would be decreased.

10.4 Database Watermarking Techniques

Basically, many watermarking techniques have been proposed for database watermarking. However, there are many factors that can be used to categorize available database watermarking techniques. These factors are including the amount of watermark distortion, type of the host data, and watermark's types. Here, available database watermarking techniques are divided into two main types including distortion-free and distortion-based.

10.4.1 Distortion-Based Watermarking

In this type of database watermarking, a watermark is embedded into the data which is caused a distortion in the database. However, the amount of data manipulation for watermarking should be small enough that data is usable and the amount of distortion is tolerable. Therefore, the watermark can be embedded at three main levels including higher level (attributes, tuples, and records), character level, and bit level. In the following, some of the database watermarking techniques for each level are discussed.

10.4.1.1 Watermarking Based on Numerical Data-Type Attribute

This technique, known as AHK algorithm, has been proposed by Agrawal et al. [2, 3] to embed the watermark at bit level of the numeric data-type attributes. This technique is created a bit pattern for a relation with specific values which performs by modifying some bit positions in some attributes in some of the tuples. For this purpose, number of candidate tuples for watermarking (γ), number of available attributes for watermarking (ν), number of available LSB in an attribute for watermarking (ξ), and secret key (K) should be applied which are only known by the relation's owner. On the other hands, these four parameters are determined bit values and bit positions algorithmically for some attributes within some tuples. For instance, the candidate bit positions can be determined by the cryptographic MAC function [4] as expressed in Eq. (10.1):

$$H(K||H(K||r.P))\qquad\qquad(10.1)$$

where P is a primary key, r is a tuple, $||$ corresponding to concatenation operation, and $H(.)$ is a one-way HASH function which decreases the probability for collision. Instead of MAC and HASH functions, another technique has been applied linear feedback shift register (LFSR) to generate pseudorandom sequence for determining the watermark positions and watermark bits [1–3]. It should be notice that robustness and security of this technique is relied on owner to protect privacy of these parameters. So far using primary key is basic assumption, but another technique is presented without applying any primary key [5]. In this technique, three major approaches are applied to obtain primary key for a relation virtually.

The probabilistic and blind nature of the watermark extraction process is based on a preselected value (α) for determining significance level of the test and the actual number of the watermarked tuples (τ). If the least τ tuples are matching with watermark pattern, the relation is considered as fake. Because the order of attributes must be fixed for watermarking extraction, the main limitation of this watermarking technique is lack of robustness against re-sort attacks which makes the detection almost impossible.

10.4.1.2 Watermarking Based on Non-numeric Multi-word Attributes

Instead of embedding watermark at a bit level of numeric attribute, there is possibility to embed watermark in non-numeric and multi-word attributes of some tuples [6]. For this purpose, the database is divided into M non-intersecting tuples for a watermark with size of $M \times N$ bits. Then, each M tuple of the database should carry M watermark strings by creating a double space for multi-word attributes of ith tuple in that subset. For watermark extraction, the embedded short binary string is extracted based on counting the number of single spaces appearing before double space. Then, these extracted binary strings are converted to the decimal equivalent. Because all non-intersecting subsets of the database are carried same watermark, this technique is robust against subset selection, subset alteration, subset addition, and subset deletion attacks. In addition, multiple small watermarks can be embedded which provide large bit capacity for this watermarking technique. However, this technique is suffering from removal attack which can be replaced all double spaces between two words (if exist) by single space for all tuples in the relation.

10.4.1.3 Watermarking Based on Tuple or Attribute Insertion

This watermarking technique is developed for unsecure communication channels where tamper proofing of the database is required. Because of the fragility nature of this technique, it is easy to detect any alternation, insertion, and deletion attacks. For this aim, two techniques have been developed which are based on inserting a virtual attribute or inserting fake tuples. The basic idea in the first technique is inserting virtual attributes for each non-overlapping partitions of a relation [7]. This virtual attribute works as a parity checksum to aggregate value obtained from any one of the numeric attribute of all tuples. However, the basic idea for second technique is generating fake tuples for inserting them into the database [8]. These fake tuples are generated based on candidate key attributes without any sensitivity on other attributes. For embedding the watermark, ith non-candidate attribute (A_i) of the relation is selected uniformly (or based on higher occurrence frequency) and Bernoulli sampling probability (Pi) is used to decide about whether it is fake value or not. For extracting the watermark which is not based on inverse process, the inserted fake tuples are checked with primary key to find similar or identical tuples for matching. Then, it is deciding whether the fake tuples inserted during watermark insertion phase or not. Although database's owner can determine the number of fake tuples which can be verified for several times and robust against incremental updatability, it is not robust against benign deletions when all of the fake tuples are deleted.

10.4.2 Distortion-Free Watermarking

In distortion-free watermarking techniques, the main aim is constructing a watermark without inserting any distortion in embedding process. Thus, embedding process is not depended on specific type of attribute. Majority of distortion-free watermarking has fragile nature which providing data integrity for the database.

10.4.2.1   Extracting Hash Value as Watermark Information

This watermarking technique has been developed for categorical data which cannot tolerate any watermark's distortion. Because of the fragility nature, this watermarking technique is applying for a database relation to detect any data manipulation. A technique has been proposed to apply primary and secret keys with hash value parameterized to partition the tuples [9]. For each pair of tuple, a watermark is extracted by using group-level hash value with same length of tuple pairs. Then, the watermark bit is embedded into the order of two tuples and based on their tuple hash values. As a result, any data manipulation is affected both groups as the manipulated tuple may be deleted from one group and be inserted to another group.

Another watermarking technique has been developed which has same idea but using categorical attribute value to partition tuples into different groups [10]. Then, group-level hash values and tuple level are computed for each group. For increasing the watermark payload, Myrvold and Ruskeys' linear permutation unranking algorithm can be applied to exchange the positions of tuples [11].

10.4.2.2   Combining Owner's Mark and Database Features as Watermark Information

A public authentication technique has been developed for data integrity based on database's tuples, watermark of the database's owner, and using hash function [12]. For this purpose, first, a watermark is generated with size of $\sqrt{n} \times \sqrt{n}$ where n is number of tuples in the relation. Second, MD5 hash function and XOR operation are applied on each tuple value to generate a value between 0 and 255 ($0 \leq Ci \leq 255$). Therefore, all Ci values are combined to produce a feature Cn where n is number of tuples in the database. Lastly, exclusive R is applied on Cn and W to produce a certification code ($R = Cn \oplus W$). In watermark detection process which is similar to embedding process, the certification code (R) is extracted to prove the database integrity.

10.4.2.3   Converting Database Relation into Binary Form Used as Watermark Information

The public watermarking technique has been proposed without using any secret key and without using any constraints on watermarking data type which can be Boolean, character, real numeric, and integer numeric. Moreover, this watermarking technique can be verified the data integrity several times. In this technique, a unique key (P) is generated based on owner's identity and database characteristics (such as database version and database name) for both verification and creation processes [13]. Then, a watermark (W) is generated based on the unique key (P) and the relation (R) which can be expressed as in Eq. (10.2)

$$W(P, W0, \ldots, W\gamma - 1) \quad \text{where } W0, \ldots, W\gamma - 1 \in 0, 1 \quad (10.2)$$

where η is a number of the tuples in W which is equivalent to original number of tuples in relation (R) of the database. The number of watermark bits $(\omega = \eta \times \gamma)$ is determined by η and the control parameter (γ) which can be considered the number of binary attributes in W and always is less than number of attributes in relation (R). Finally, the order of the attributes and the MSBs of the attribute values are randomized by using a cryptographic pseudorandom sequence generator (e.g., LFSR) to generate the watermark. Although the MSBs can be detected any modification due to intolerable characteristics, modification on other bits cannot be revealed in this watermarking technique.

10.4.2.4 R-Tree-Based Permutation as Watermark

In order to remove any condition on the order of entries inside the node, a watermarking technique is proposed based on the R-tree data structure [14]. Depending on the watermark value, the entries inside of R-tree nodes are reordered based on a secret key which is defined the initial order. For this purpose, a one-to-one transformation function is performed for mapping between all the watermark values and all possible permutations of entries inside of R-tree nodes. This mapping is using variable base with factorial value to construct a numbering mapping system. Therefore, these watermarking techniques provide some advantages including less distortion due to the data value in R-tree node is not modified, R-tree size is not increased, the R-tree watermarking is not interfered with R-tree operations, blindness, and minimal overhead.

10.5 Types of Digital Watermark

Different watermarking techniques have been discussed in previous part. Here, these techniques are discussed in terms of the watermark information types. Much information can be embedded into the relational database. In this section, available types of information that have been used for watermarking including arbitrary meaningless bit pattern, content characteristics, cloud model, meaningful information, fake tuples, virtual attribute, image, and speech are discussed.

10.5.1 Arbitrary Meaningless Bit Pattern

The watermark data can be generated randomly without any meaning or special pattern which makes the watermark robust against insensitive to an initial value, collision in a hash function and repetitive iterative operation. In [15], random integer numbers has been embedded into the LSB positions of some attributes inside some tuples which are selected algorithmically.

10.5.2 Image

The watermark data can be an image with size $N \times N$ which should be scrambled into a binary string with of length $L = N \times N$. In [16], L groups are created from all tuples in the relation. Then, a hash function is applied on order of the image, tuple's primary key, and secret key to determine the belongs group for each tuple. Lastly, the ith watermark bit (L_i) is embedded into the bit position of the attribute value for ith group of the tuples algorithmically.

10.5.3 Speech

The watermark data can be the speech of the database owner which requires some preprocess steps to convert the speech signal into bitstream. For this purpose, firstly, the speech should be compressed; secondly, the noise should be removed from the compressed speech by applying speech enhancement technique; thirdly, the enhanced speech signal should be transformed into bitstream; finally, a watermark is generated based on this bitstream which can be embedded into the database [16, 17].

10.5.4 Content Characteristics

The watermark data can be considered as the characteristic attribute A1 of tuple t which is called local characteristics and embedded into the watermark attribute A2 of same tuple t [18]. Depending on the non-NULL requirement of characteristic attribute value and generated random value (between 0 and 1) which should be less than α (that is embedded proportion of the relational databases), the candidate tuples are chosen. For watermark extraction, firstly, local characteristics of the characteristic attribute are detected with similar procedure in embedding process. Then, the last bits of watermark attribute are compared with them.

10.5.5 Cloud Model

Basically, cloud model has three major characteristics including hyper entropy (He), entropy (En), and expected value (Ex) which can be used for database watermarking [19]. On the other hand, it is possible to embed cloud drop as a watermark (which by using cloud generation algorithm) into relational database. In the embedding process, a cloud with parameters Ex, En, and He is generated by forward cloud generation algorithm and it is embedded in the relation of the database as a watermark. In the non-blind extraction process, the backward cloud generation algorithm is applied to detect both embedded and extracted cloud drops. Finally, similar cloud algorithm is used to verify whether both clouds are similar or not.

10.5.6 Meaningful Information

A meaningful information can be considered for watermarking which needs to transform it to bit flow. In [20], some unique IDs are computed from all the tuples of the relation. Then, these IDs are sorted in ascending order by their values. For embedding process, ith bit flow of the meaningful watermark is embedded into ith tuple group by AHK algorithm. For blind extraction process, only tuples are selected but rolled back to decrease FPR.

10.5.7 Fake Tuples

Fake tuples can be generated and embedded as a watermark data into the database. These fake tuples are generated based on candidate key attributes without any sensitivity on other attributes. In [8], the fake tuples have been applied as a watermark to provide tamper detection for watermarked database.

10.5.8 Virtual Attribute

Similar to fake tuples, a virtual attribute can be served as a watermark. Not only the virtual attribute can obtain from any of the numeric attribute of the whole tuples but also it contains parity checksum of the other attributes. In [7], virtual attribute scheme has been used for watermarking to authenticate the watermarked database over unsecure channel.

10.6 Extension of Database Watermarking to XML Watermarking

A XML database is a kind of document-oriented database system with ability of storing, querying, transforming, specifying, and returning in XML format. Nevertheless, some watermarking techniques have been extended their solution from attributers in relational databases to XML elements [21].

10.7 Preliminary in XML Data

Basically, a tree structure is used to represent a XML data as presents in Figure. Two types of nodes are constructed in the tree: leaf node and non-leaf node. In the leaf node, only a value is presented without referring to other nodes. However, in non-leaf node, a set of edges are referring to other nodes without specifying any value. In a XML schema, node name and their relationships are defined but in a XML tree, the instances of XML data are conformed to a certain schema. Other definitions are presented in the following:

Path: a string is constructed from XML nodes by considering parent–child hierarchy. For example, bookstore/book/title is a path.

Query: A query is a string that contains some paths (Pi) and some selection criteria (Si). The selection criteria (Si), which themselves contain subpaths and values, must be satisfied to give some nodes in Pi. For example, a query is bookstore/book [category = "Web"]/title.

Query Template: Query template is a query without comparison operations of the query. For example, a query template is bookstore/book [category]/title.

Tree Tuple: A tree tuple is a set of nodes in a non-recursive XML tree that can be reached by path and parent–child hierarchy is preserved.

Flat Table: A flat table is another representation of a XML tree in which each column corresponds to a path rooted at the root, and each tuple corresponds to a tree tuple.

Functional dependency: a functional dependency is an expression of the form $P1 \rightarrow P2$, where P1 and P2 are member of paths (S). T satisfies $P1 \rightarrow P2$, if for any pair of tuples d1; d2 in the flat table FT (T), $d1.P1 = d2.P1$ implies $d1.P2 = d2.P2$.

10.8 XML Watermarking

Today, XML is the most common data representation for exchange over Internet networks. The popularity of the XML data for exchanging and distribution can cause unauthorized duplication by different Internet parties and applications. For instance, there is possibility to steal some information such as a job advertisement or information about book's collection in the formant of XML for a job agent and digital library Web sites. The information can be easily reused in other Web site. Therefore, a mechanism is required to protect the XML ownership which can be solved by using digital watermarking technology. On the other word, digital watermarking technology can prevent reproduction of the XML data by embedding indiscernible watermark. XML watermark may have challenging tasks due to its structure and data elements which are not isolated and they may be linked by a relationship. This fact can reduce the embedding capacity for XML watermarking.

A XML data can be modeled by a tree structure (as shows in Figure) which provides two main types of semantics information including functional and keys

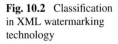
Fig. 10.2 Classification in XML watermarking technology

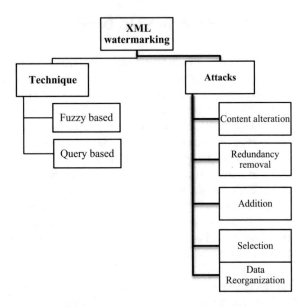

dependencies. As seen in figure, title of a book can be considered as key due to it is unique. The functional dependency is defined by relationship between an author and a book. Both functional and key dependencies contain crucial XML data relationship which may compose XML data redundancy. These key and functional dependencies construct identifiers (which are independent from data redundancies and can differentiate different data elements) for XML watermarking. In order to improve resilience and data usability of the XML watermarking technique, the query templates are also used for production of the identifier [22] (Fig. 10.2).

The queries for identifying data elements and structure units must have three criteria. Firstly, the queries should have ability to differentiate among various data element to provide enough watermarking payload. Secondly, they should be able to defend against removal attacks by identifying redundancy among XML data. Thirdly, they should be able to survive under alteration and reorganization attack by providing data usability. In the following, two major XML query-based watermarking techniques are discussed.

10.8.1 Query-Based XML Watermarking

To overcome the XML challenges, some schemes such as graph labeling, semantics, and queries can be used for XML watermarking [23]. However, query is an adaptive and simple identifier of data elements and structure units which is independent from physical organization of the XML data. The data usability is represented by query templates that are generated by semantics in order to identify structure units and data elements in the XML content for watermarking. Three main steps in XML watermarking are including initialization, watermark embedding, and watermark extraction which are discussed separately as illustrates in Figure.

(a) Initialization: In this step, a schema is specified and validated the XML data. Also, a set of query template (Q) is generated to represent data usability. A number of structure units and data elements for embedding a watermark are selected based on a secret key. These structure units and data elements are identified to safeguard in the query set (Q) by creating some queries.

(b) Watermark embedding: In this step, the structure units and data elements are retrieved by executing the queries in the query set (Q). Then, a watermark embedding algorithm is selected to insert the watermark bits into these elements and units.

(c) Watermark extraction: In this step, the same queries in the query set (Q) in the watermark embedding step are executed to retrieve the structure units and data elements. However, these queries should be rewritten based on the transformation from the original schema to a new schema due to reorganization attack.

10.8.2 Fuzzy Queries XML Watermarking

Fuzzy queries XML watermarking is another kind of query-based watermarking technique that applies fuzzy queries to retrieve structure units from XML data with some approximation [24]. Basically, many criterions should be considered to choose the locators. This watermarking technique is solved the problem of extracting locators of the XML data by using fuzzy queries. The basic idea is that the tag for an XML data which is mostly used by users and services. On the other hands, the tag is extracted from a XML data by using more frequent executed queries. After the locator selection process, the owner pattern (as watermark) is embedded into these locator by assigning a fuzzy query for each locator. Using fuzzy queries makes this watermarking technique more robust against deletion attacks (Fig. 10.3).

10.9 XML Watermarking Attacks

There are different attacks which can degrade the watermarked XML data. These attacks can be classified into data reorganization, redundancy removal, addition, selection, content alternation attacks. In the following, each attack is discussed in detail.

(a)

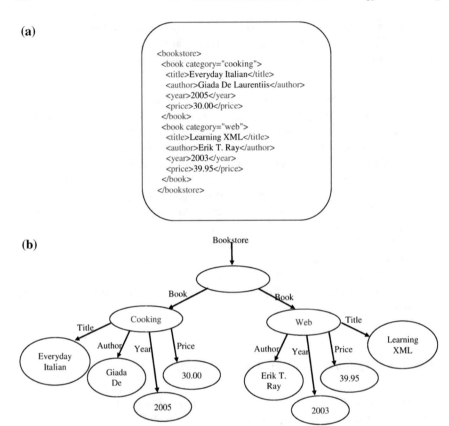

(b)

Fig. 10.3   **a** XML code **b** XML tree

Data Reorganization attack:

In order to disturb the owner for extracting watermark which has already embedded into the structure units and data elements of the watermarked XML data, the watermarked XML data is redesigned and reorganized based on a new schema. This attack is mainly applied for XML data due to data elements and the relationships in XML data are much more alterable and flexible than other data such as audio, video, and image.

Redundancy Removal attack: Generally, redundancies within the XML data are added intentionally by semantics to provide simplicity and performance. Some of the XML watermarking techniques have developed their schemes based on embedding in redundant XML data. Similar to compression attack for multimedia that removes redundant data, redundancy removal attack is exploited to destroy the redundancies within the XML data which can deter the watermark extraction. This attack can seriously degrade the watermarked XML data which has embedded in the duplicates of the XML. Although there is normalization technique for removing any redundancy from XML data before it is watermarked, the XML data normalization technique is not efficient enough still is under research [25]. Moreover,

Addition Attack: For addition attack, the watermarked XML data is mixed with other related XML data in order to prevent watermark to be extracted.

Selection Attack: This attack selected a specific part of the watermarked XML data which can satisfy the attacker intend for usage. Thus, other parts of the watermarked XML data are discarded in a hope that watermark has already embedded in these parts.

Content Alteration Attack: Content alteration attack is the most common attack with aim to degrade the watermark extraction by modifying the structures and elements of the watermarked XML data selectively or randomly. However, this attack is limited to data usability factor which should not be seriously degraded.

10.10 Extension of Database Watermarking to Ontology Watermarking

Relational database watermarking techniques can be extended to ontologies watermarking which alter data for Semantic Web [26]. Although this data alternation may degrade the ontology's precision before publishing, this data alteration has less degradation than malicious republication scenario due to incompleteness in the nature of the most ontologies.

10.11 Ontology Watermarking

A formal collection of world knowledge is known as ontology. Creation of the ontology requires knowledge collection, formalization and maintains processes which can be constructed by human effort or machine (automatically). Because creating an ontology requires huge amount of time and cost for algorithms development and scientific investigation, ontology usually is not provided for arbitrary purpose and for free. Therefore, owners of the ontology use a specific license to prohibit republish the ontology [26, 27]. Even the public ontology of the Semantic Web cannot be republished arbitrary, and any redissemination of the data is considered as a dishonest [28]. The main concept behind the ontology of the Semantic Web is sharing data across community [29]. However, detecting republication of an ontology illegally is the major issue for owners. Although there is possibility to upload an ontology to trusted external server, an adversary can claim that s/he collected these ontologies herself/himself before uploaded by real owner. Therefore, a sophisticated solution like digital watermarking is critical to prove the ownership claim for these ontologies. Embedding some relevant information into ontologies

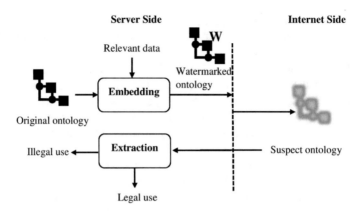

Fig. 10.4 Ontology watermarking concept for proof of ownership

in invisible and robust ways can protect the ownership against republication operation (Fig. 10.4).

References

1. Halder, R., S. Pal, and A. Cortesi. 2010. Watermarking techniques for relational databases: Survey, classification and comparison. *Journal of UCS* 16(21): 3164–3190.
2. Agrawal, R., P.J. Haas, and J. Kiernan. 2003. A system for watermarking relational databases. In *Proceedings of the 2003 ACM SIGMOD international conference on management of data*. ACM.
3. Agrawal, R., P.J. Haas, and J. Kiernan. 2003. Watermarking relational data: Framework, algorithms and analysis. *The VLDB Journal* 12(2): 157–169.
4. Agrawal, R., and J. Kiernan, 2002. Watermarking relational databases. In *Proceedings of the 28th international conference on Very Large Data Bases* (pp. 155–166). VLDB Endowment.
5. Li, Y., V. Swarup, and S. Jajodia. 2003. *Constructing a virtual primary key for fingerprinting relational data*, 133–141. Workshop: In Digital Rights Management.
6. Al-Haj, A., and A. Odeh. 2008. *Robust and blind watermarking of relational database systems*.
7. Prasannakumari, V. 2009. A robust tamperproof watermarking for data integrity in relational databases. *Research Journal of Information Technology* 1(3): 115–121.
8. Pournaghshband, V. 2008. A new watermarking approach for relational data. In *Proceedings of the 46th annual southeast regional conference on XX*. ACM.
9. Li, Y., H. Guo, and S. Jajodia. 2004. Tamper detection and localization for categorical data using fragile watermarks. In *Proceedings of the 4th ACM workshop on digital rights management*. ACM.
10. Bhattacharya, S., and A. Cortesi. 2009. A distortion free watermark framework for relational databases. In *ICSOFT (2)*.
11. Li, Y. 2008. Database watermarking: A systematic view. In *Handbook of database security*, 329–355. Springer.
12. Tsai, M.-H., et al. 2007. Fragile database watermarking for malicious tamper detection using support vector regression. In *Third international conference on intelligent information hiding and multimedia signal processing, 2007. IIHMSP 2007*. IEEE.

13. Li, Y., and R.H. Deng. 2006. Publicly verifiable ownership protection for relational databases. In *Proceedings of the 2006 ACM symposium on information, computer and communications security*. ACM.
14. Kamel, I. 2009. A schema for protecting the integrity of databases. *Computers & Security*, 28(7): 698–709.
15. Qin, Z., et al. 2006. Watermark based copyright protection of outsourced database. In *10th International on database engineering and applications symposium, 2006. IDEAS'06*. IEEE.
16. Wang, C., et al. 2008. Atbam: An arnold transform based method on watermarking relational data. In *2008 International conference on multimedia and ubiquitous engineering*. IEEE.
17. Hu, Z., Z. Cao, and J. Sun. 2009. An image based algorithm for watermarking relational databases. In *International conference on measuring technology and mechatronics automation, 2009. ICMTMA'09*. IEEE.
18. Zhang, Y., et al. 2006. *Relational databases watermark technique based on content characteristic*. In *First international conference on innovative computing, information and control, 2006. ICICIC'06*. IEEE.
19. Zhang, Y., X. Niu, and D. Zhao. 2005. A method of protecting relational databases copyright with cloud watermark. *International Journal of Information and Communication Engineering* 1(7): 337–341.
20. Huang, M., et al. 2004. A new watermark mechanism for relational data. In *Null*. IEEE.
21. Zhou, X., et al. 2005. *Wmxml: A system for watermarking xml data*. In *Proceedings of the 31st international conference on very large data bases*. VLDB Endowment.
22. Rane, M.P.B., and B. Meshram. *Xml-Based security for e-commerce application*.
23. Gross-Amblard, D. 2011. Query-preserving watermarking of relational databases and xml documents. *ACM Transactions on Database Systems (TODS)* 36(1): 3.
24. Romaric, T., E. Damiani, and N. Bennani. 2012. Robust XML watermarking using fuzzy queries. In *2012 IEEE 36th annual on computer software and applications conference workshops (COMPSACW)*. IEEE.
25. Mat Kiah, M., et al. 2011. A review of audio based steganography and digital watermarking. *International Journal of Physical Sciences*. 6(16): 3837–3850.
26. Franco-Contreras, J., et al. 2014. Ontology-guided distortion control for robust-lossless database watermarking: Application to inpatient hospital stay records. In *2014 36th Annual International Conference of the IEEE Engineering in Medicine and Biology Society (EMBC)*. IEEE.
27. Suchanek, F.M., D. Gross-Amblard, and S. Abiteboul. 2011. Watermarking for ontologies. In *The semantic web–ISWC 2011*, 697–713. Springer.
28. Pazienza, M.T., A. Stellato, and A. Turbati. 2008. Linguistic watermark 3.0: An RDF framework and a software library for bridging language and ontologies in the semantic web. In *SWAP*.
29. Kong, H., et al. 2009. Techniques for owl-based ontology watermarking. In *WRI global congress on intelligent systems, 2009. GCIS'09*. IEEE.

Part IV
Advance in Watermarking

Chapter 11
Network Stream Watermarking

11.1 Introduction

Detection of stepping stones is a famous problem in computer security. In order to hide the identity, Internet attackers broadcast their traffic on multiple hosts known as stepping stones. A common approach to detect these hosts is to observe the pattern of hosts' entering and leaving the network. Traditionally, the pattern inherent of packet timing, sizes, and counts is utilized in traffic flows for the analysis of the incoming and outgoing flow of a host [1].

Intrusion detection suffers from the linking network flows, as well as anonymity problem. From a point of view, the techniques for the analysis of network traffic are classified into passive and active. Although passive techniques are able to link flows, a high time complexity is required for low rate of errors. On the other hand, active techniques are more precise and scalable [2, 3]. Active techniques or flow watermarking disconcerted the characteristics of traffic on an incoming flow once they pass routers to generate a distinct pattern. This is detectable in outgoing flows. Active techniques can break the anonymity when two flows are linked. Application of active and passive techniques in anonymous communications has been widely studied in the literature [2–4].

Passive and active techniques demonstrate some features that may compromise one another in traffic analysis applications. For long-lived network flows, passive techniques exhibit good performance, but a large amount of data on traffic flows are required. Meanwhile, flow watermarking observes short period of traffic flows efficiently. Flow watermarking is considered as blind watermarking because the required information for flow watermarking is carried by the flow. Therefore, additional storing and communication of network flows as needed for passive techniques is not essential. Although blind watermarking is more robust, such robustness imposes extra cost to the watermarking technique. It is a consequence of large delays (hundreds of milliseconds) to the flows and makes the benign users susceptible to attacks [5].

© Springer Science+Business Media Singapore 2017
M.A. Nematollahi et al., *Digital Watermarking*, Springer Topics
in Signal Processing 11, DOI 10.1007/978-981-10-2095-7_11

Because of a trade-off between passive and blind flow watermarking, non-blind flow watermarking was proposed. Non-blind flow watermarking records the pattern of incoming traffic flows and associates them to the outgoing flows that is similar to passive techniques. However, they change the communication patterns of the intercepted flows which are similar to the blind flow watermarking techniques.

A prototype for recording the timing pattern of incoming flows of non-blind flow watermarking and associating them with timing pattern of outgoing flows is RAINBOW. RAINBOW adds a watermark to each incoming flow by delaying some packets once the received timings are recorded. For shortening the delays, RAINBOW utilizes the benefits of spread spectrum. As the watermark is not related to the flows, this watermarking technique reduces the effect of natural similarities between two unrelated flows and the required time for flow linking decision.

As the delays are limited to milliseconds, no interference with the traffic of normal users is occurred and the watermark result becomes invisible. Moreover, the natural network jitter is in the same magnitude of watermark delays, and the probability of watermark visibility by statistical analysis gets very low [6, 7]. The invisibility of RAINBOW has been approved by using an information theoretical tool [7]. It also showed high performance of RAINBOW in linking network flows through a prototype implementation over the PlanetLab [8] infrastructure.

11.2 Network Traffic Modeling

Network watermarking techniques can be analyzed over a network traffic model. As the network traffic depends on many parameters, the presentation of a comprehensive network traffic model that reflects the real network traffic is very complicated. So far two models for network traffic have been introduced and any real-world traffic model falls between these two models.

The first model observes each traffic flow as an independent flow that is demonstrated as a Poisson process [each flow has, i.e., inter-packet delays (IPDs)]. It is suitable for non-interactive network flows. Completely correlated flows (with similar timing patterns) are taken into account by the second model. Correlated IPDs are the case where all the network flows have the same IPDs. Common traffic types including browsing the same/similar Web sites, video streaming, bulk file transfers, and VoIP voice/video calls are captured well by this model. In general, watermarking is more efficient than passive traffic analysis in linking this type of traffic.

11.3 Network Watermark Properties

A proper network watermark should fulfill some requirements. Robustness of the network watermark against the occurrence of the traffic modifications (e.g., jitter) is among the well-known properties of a network watermarking technique. A desired watermark technique has the lowest distortion on the performance of the flows, especially for stepping-stone scenarios where watermark flows are almost benign. Invisibility of watermark is another important property expected from network watermark techniques. Invisibility implies the watermark to remain invisible to attackers.

Generally, robustness compromises the invisibility of watermark. A watermark remains invisible if it makes the least modifications to the packet stream. If the packet delays in watermarking exceed few milliseconds, they are detected from natural network jitters. Therefore, watermark loses the invisibility.

Timing-based watermarking techniques are counted as active traffic analysis methods. They provide a high precision and robustness by efficient linking network flows that are dispatched across one or more relay hosts. When distortion noises are existed in the network, e.g., network jitter, low values of FNR and FPR can be achieved by watermark detection. However, the detection of the watermark is not always possible in large networks, even when the key is available.

If the watermark tolerates natural timing distortion, a low rate of FNR is expected. As watermark detection is sometimes challenging, in some cases multiple copies of the watermark are embedded in different positions of the flows. Therefore, timing distortions cannot destroy the whole watermarks and a low FNR is obtained. Packet reordering, packetizing, delay by congestion, and packet loss are instances of timing distortions on network flows. FNR becomes high when a determined active adversary is faced. It is due to removing the timing-based watermarks through drastic measures.

One of the methods used by active attackers to remove or alter timing-based watermarks is adding some dummy packets, sending packets in batches, and/ or introducing large delays. Fortunately, these methods impose an acceptable cost to attackers especially in real-time applications, e.g., Tor and SSH stepping stones. Consequently, active attackers prefer to emerge passive and reactive launch counterwatermark attacks. Passive invisibility is an outcome of low FNR against adversaries. In general, active countermeasures are not useful when watermark is invisible to passive attackers. Nevertheless, low FNR can compromise the invisibility of the watermark. When more network watermarks are hided over the flows, the probability of their picking by the attackers gets higher.

Some parts of the signal may be selected false as the watermark portions. Low FPR denotes that the number of false picked marks is the minimum. High FPR reduces the efficiency of watermark over passive traffic analysis. As the high rate of FPR misleads the attackers, making it low against adversarial manipulation is not very serious. Although the attackers can embed multiple copies of the watermark to confuse the detector [5, 6], it is hard to blind-embed the benign flows without having the secret key. A method for increasing FPR by embedding false marks and no secret key is the best-effort copy attack. An application of copy attacks by replicating the watermark on outgoing circuits and increasing the FPR is in busy Tor relays.

11.4 Network Watermarking Techniques

Various techniques have been developed for network flow watermarking with different watermarking strategies. In addition, various adversaries can attack in different ways. To provide an overview of network stream watermarking technology, Fig. 11.1 shows all of these issues. In the following, each of these issues is discussed separately.

11.4.1 Network Flow Watermarking

Network flow watermarking employs the watermark to network flows in a similar way to other digital watermarking techniques. Generally, a secret key is utilized for embedding the watermark and extraction of the watermark from network flows. The network flows traversed from a watermarking point that is normally a router. The watermarking point alters or transforms the flow. Packet timing, which delays the packets, is the common method for flow alteration.

For more robustness, the network flows are altered by naturally or intentionally distortions. Instances of natural distortion are the delay in the intermediate routers or other variables of delays, i.e., jitter, or modifications including repacketizing, dropped packets, or retransmitted packets. Alteration of network flow after watermarking may also be the result of attacker's distortions on the traffic

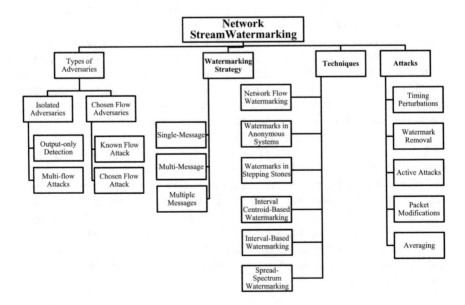

Fig. 11.1 An overview on network stream watermarking

characteristics of flows, in order to make the signal unrecoverable by the encoders. The distorted signal finally reaches to a detection point where the secret keys are applied to extract the watermarked message. A proper watermarking technique should allow the watermark message to be read at the destination even though it is distorted despite distortions by the routers and the attackers.

Network flow watermark techniques can send a single message by watermarked flows or different messages by different flows. In the first case, the detector realizes the presence or absence of the watermark in the received flows by checking the correctness of the decoded message. In the second case, a detected watermarked flow is linked to another marked flow. However, it reduces the probability of reliable detection of the watermark. Generally, sending a single message improves the error detection of network flow watermarking.

11.4.2 Watermarks in Anonymous Systems

Anonymous system links the input traffic flows to some output flows of the network, but hides the pattern of their connection. Therefore, an attacker may alter the pattern of connections for distorting the watermark. In order to implement the pattern, an onion routing [9], a mix network [10], or a simple proxy [11] can be utilized. Anonymity protection benefits watermarks through embedding marks in certain input flows. Then, the watermark is extracted by inspecting the output flows. An attacker can steal data on an anonymizing system after the user made connection to a malicious Web site. For this purpose, a coattacker helps the attacker to snoop on the links between anonymous system and user to find out whether the user is browsing the site. In another situation, all the flows going from a compromised entry router in Tor [12] can be watermarked. Then, the watermark can be revealed by a cooperating exit router or a Web site.

If the flows at two points of low-latency anonymous systems are observable by an attacker, he/she can detect the sameness of them, unless a cover traffic is utilized [13]. Due to the cost of using cover traffic, it is not usually deployed in real low-latency systems, hoping that the attacker is not able to detect a noteworthy part of the input and output flows. This strategy has been used in Freedom [14], onion routing [9], and Tor [12].

When passive traffic analysis is utilized and n input flows and m output flows are detected by two separate attackers, $O(n)$ communications between the attackers and $O(nm)$ computation are required. In this case, the characteristics of all n flows must be transferred from an attacker to the other one. Then, each output flow must be corresponded to an input flow. From the other hand, when watermarking is applied, there is no need for further communications between the two attackers; instead, they use a secret key. In this case, the computation cost at the watermarker and detector reaches to $O(n)$ and $O(m)$, respectively.

11.4.3 Watermarks in Stepping Stones

When there is a remote connection between two parts of the network, an intermediate host known as stepping stone acts as the role of a traffic relay to hide
the origin of the network flow. Stepping stones can be detected by the initiative
host if the initiative host links an incoming flow to the relayed outgoing flow. This
case is similar to an anonymous communication system with n incoming and m
outgoing flows. The cost of this task by using watermarking is lower than passive traffic analysis [15, 16]. The computation cost of passive techniques is $O(nm)$
and reaches to $O(n)$ for communication when the traffic flow enters and leaves the
enterprise via multiple border routers. The responsibility of the border routers for
an enterprise is to embed the watermarks on all incoming flows and to detect the
presence of the marks on all outgoing flows. This process reduces the computation
cost to $O(n)$ for incoming flows and to $O(m)$ for outgoing flows.

11.4.4 Interval Centroid-Based Watermarking

Equal length intervals are the result of dividing the stream in interval centroid-
based watermarking (ICBW) [17]. For this purpose, two parameters including
the offset of the first interval, o, and the length of each interval, T, are employed.
$2n = 2rl$ intervals are randomly selected to form a set. This set later breaks into
two sets A and B, each containing $n = rl$ intervals, in a random fashion. A and
B are, respectively, divided to L subsets $\{A_i\}_{i=1}^L$ and $\{B_i\}_{i=1}^L$, each containing of r
intervals. Ai and Bi represent the sets as used for encoding the ith watermark bit.
This process embeds a watermark of length 1 in the traffic flow.

   The parameters o and T are known by watermarker and detector. A random
number generator (RNG) and a random value for seed, s, are utilized to select
and assign the candidate intervals for embedding the watermark bits. These values should be kept secret to retain the transparency (imperceptibility) of the watermark. The arrival time of the packets is delayed by the watermarker, regarding the
value of the ith watermark bit (either 1 or 0). The delay is applied in set Ai or Bi
at the interval positions by maximum value of a (where a is the distance between
the aggregate centroid of Ai and Bi). The delaying strategy over the distribution of
packet interval times in an interval of size T is known as squeezing. The offset of
packet arrival time in average represents the aggregate centroid value. It measures
the length T from the beginning of the current interval. The corresponding interval
to ith bit is increased by $\frac{a}{2}$. When the watermark bit is 1; the intervals Ai and Bi are
the corresponding intervals. Therefore, the aggregate centroid of Ai and Bi has the
difference of $\frac{a}{2}$ when the watermark bit is 1 and $-\frac{a}{2}$ when the watermark bit is 0.

   Realize that watermark bit is performed by the detector. In order to detect the
value of the watermark bit, the detector calculates the difference between the

aggregate centroid of intervals Ai and Bi. When the result is closer to $\frac{a}{2}$, the watermark bit is detected as 1. On the other hand, when the result is closer to $-\frac{a}{2}$, the watermark bit is detected as 0.

The robustness of ICBW against insertion of chaff, repacketizing, and mixing of data flows is high. It is the result of using r arrival times of many intervals for each watermark bit. If the values of a and T are considered large enough (i.e., 350 and 500 ms), network jitter can solely shift few packets from an interval to one another. As the interval positions of Ai and Bi are secret and the generated patterns by a natural variation are not simply realized in high traffic from a watermark bit, detection or removing watermark bits is hard in ICBW. Nevertheless, a simple and effective approach recovers the watermark bit and its position. It can also be used for any watermark technique that benefits the generation of intervals on specific fragments of the traffic flow across many flows.

11.4.5   Interval-Based Watermarking

As an extension of ICBW, interval-based watermarking (IBW) [18] employs the arrival times of packets over specific intervals. The traffic rates are then manipulated in successive intervals for inserting the watermark bits. This method utilizes two different intervals for embedding the watermark; however, both intervals are manipulated during watermarking. All packets between intervals I_i and I_{i+1} are delayed to generate a clear area, and all the packets between intervals I_{i-1} to I_i are delayed for loading the watermark, or vice versa. By using this strategy, the number of packets in loaded interval becomes twice, while the clear interval has no packets. If the watermark bit is 0, the loaded interval is I_{i+1} and the interval I_i is cleared. Otherwise, I_i is loaded and I_{i+1} is cleared for sending bit 1. Three intervals are taken for sending a bit, because one interval is cleared and it implicitly loads the next.

In addition to the parameters o and T, a list of positions for embedding the watermark $S = \{s1,\ldots, sn\}$ is kept secret by the watermarker and the detector. The presence of watermark bits is checked by analyzing the difference between the data rate in interval I_{si} and interval I_{si+1}. A predefined threshold is utilized for detecting this difference. If the intervals are different, a watermark bit is detected. The detector suffers from the problem of false positives and false negatives. It is the result of natural variation in packet rates as false presence of the watermark is detected (false positive). Due to the delays between the watermarker and repacketizing at the relay, the rates in intervals may shift and false-negative problem may occur. However, the rate of false positive and false negative is very low. This approach is robust against repacketizing and encodes each watermark bit in several positions to improve the reliability of transmissions.

11.4.6 Spread-Spectrum Watermarking

Spread-spectrum watermarking [19] embeds the watermark bits in the traffic flow for invisibility of the trace back. By taking n as the length of a binary watermark, this approach inserts the watermark through an interval of length Ts. Therefore, the total length of the watermark is nTs. The packet rate of a specific interval (with length Ts) is manipulated by pseudo-noise (PN) code for embedding bit 1. PN code switches the signal between $+1$ and -1 fastly. Duration of each ± 1 period is Tc. The flow rate is not altered when PN code is $+1$; however, it degrades for $Tc.2$ period when PN code is -1. The PN code as used in [19] is length-7. The implementation of bit 0 is similar to bit 1, by using the complement of the PN code. An instance of embedding the watermark 110 for a length-5 PN code is shown in Fig. 11.2.

The PN code and Ts are known by the detector and watermarker. At destination, a high-pass filter is applied on the signal for the extraction of the watermark. Afterward, the signal dispreads and passes through a low-pass filter. In the watermark techniques like DSS, the flow rate over certain intervals is reduced across all flows which make the watermark subjected to multi-flow attacks (MFA).

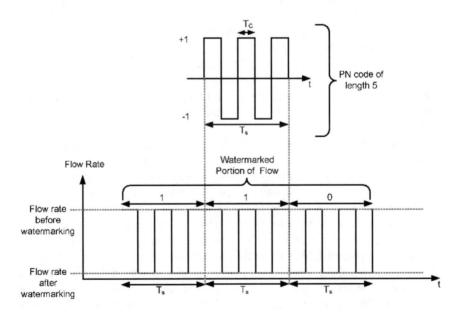

Fig. 11.2 Spread-spectrum watermarking for network stream [5]

11.5 Network Watermarking Attacks

There are various types of attacks that can threat network stream [6]. In the following, each of them is discussed.

11.5.1 Active Attack

Active attacks impose large delays to the signal by which the watermark is probably wiped out. This delay can be 500 ms. Additional measures might be applied in order to hide the attacker's tracks, e.g., dummy packet insertion which disarms any linking technique by an independent Poisson distribution [20].

11.5.2 Copy Attack

A copy attack attempts to mislead the detector between the flows. An attacker can copy a specific portion of a multimedia (over a network) to another multimedia from the same type to modify the verification process of the ownership or the authorship. Protocol attacks on multimedia watermark have been introduced in [21]. Duplicate attack as presented in [22] can be counted as a copy attack. So far, copy attacks on network flow with no knowledge on watermark parameters have not been studied.

11.5.3 Ambiguity Attacks

An ambiguity attack tries to doubt about reliability of the decision in watermark detector. Consequently, this attack embeds multiple watermarks to extract random watermark subsets at the detector in order to increase FPR and FNR. In the context of an ownership, ambiguity attack increases the FPR at watermark embedding/extractor processes. On the other word, this attack even can threat the robust watermarking techniques by obsession about possibility of the erroneous decisions. Three major groups of ambiguity attack are inversion attack, non-invertible attack, and intervention of trusted party (potential attack) [23]. Inversion attack is based on invertibility of embedding process which applies brute-force search to find a fake watermark in watermarked object. Non-invertible attack applies diffusion property to generate many watermarks from a severely altered but imperceptible version of the watermarked object to increase FPR. Potential attack is similar to non-invertible attack, but a fake original stream at a trusted party is required by an adversary.

11.5.4 Packet Modification Attack

Packet modification attack modifies the data transferred over network. For this purpose, an attacker adds or removes some packets. A practical watermark detector should resist packet additions and removals.

11.6 Watermarking Strategy

The strategy of watermark discusses the embedding technique over network flows [5]. Multiple messages can be used for different network flows or a single message can be utilized across all interactive flows. Regardless of the strategy, watermarks are applied by clearing the same parts across various flows. In some cases, each watermark uses different messages. The strategies of watermarking are discussed as follows.

11.6.1 Single-Message Watermarks

The same and single message is passed across all interactive flows in single-message watermarks. As the main problem of this strategy in the sorted union of packets' arrival times, the revelation of the kth watermarked flow can form an aggregate of all the flows.

11.6.2 Multi-message Watermarks

Different messages on different network flows are utilized in multi-message watermarks. Therefore, simple aggregation (as used in single-message watermarks) cannot be used for encoding the messages. Due to multiple switches between 1 and 0 in different messages, different transforms are applied on different intervals.

As an instance, consider ICBW where it squeezes a given interval for bit 0 and does not squeeze when it is 1. In this case, there is no empty period when flows are aggregated and bit changes occur.

11.6.3 Multiple Messages

Each watermark can utilize different messages, despite the case where the same watermarks are sent over all aggregated flows. The major drawback of this strategy is with the recovery of the secret parameters. When there are subsets of the flow, the positions of large intervals are not accurately aligned. If the interval is squeezed for 8 or 9 of the flows in the subset, and a few packets are added to the last one or two flows, the size of intervals might be detected incorrectly. One solution for this problem is to choose the largest intervals in any subset, i.e., the squeezed intervals on every flow. By using these intervals, the recovery of a and T parameters is possible. There are some intervals that continuously squeezed at the same time which correspond to either Ab or Bb for some bit b. The next step is to find these intervals by scanning all subsets and labeling them with Si. If for each Si, there is a Sj that is never squeezed simultaneously with Si, and then, Si and Sj are related to the same bit. Therefore, the watermark can be removed by observing the watermarked stream. For this purpose, the intervals from Si are artificially squeezed or unsqueezed or both.

When all the flows are not watermarked, subset technique is also applicable. Consider a Web site that embeds watermark onto certain connections. In this case, the watermark can be recovered when all watermarked subsets are recognized. In general, a watermark scheme that uses multiple messages simultaneously and embeds the watermark based on statistical methods is a challenge to the attackers. As all the flows are allowed to be marked, these methods are desired for most applications.

11.7 Types of Adversaries

An adversary against watermarking can be a stepping-stone intruder or anonymity system relay. The aim of an adversary against watermarking is to control the host that passes the traffic of network from encoder to decoder. Two classes of treat models are shown in Fig. 11.3 [6]. As seen, for isolate adversary, an adversary has access to both output of the watermark embedding and input of the watermark extraction processes. For chosen flow adversary, an adversary has access not only to both the watermark embedding and input of the watermark extraction processes but also to input of the watermark embedding process.

They are isolated adversary and chosen flow adversaries. Two further classes for both treat models can be assumed including passive (i.e., pure observers) and active (i.e., traffic manipulators). Tor is an instance of isolated active adversaries. Although Tor can influence packet timing by its relays, this action rarely occurs. It is due to the purpose of Tor relays to forward the traffic in the shortest possible time. However, Tor relays are able to change the watermarking in an active manner. Unless a high delay is not happened by changing the packet timing, Tor relays inject delays to the watermark. They even may exchange watermarks between circuits.

An adversary can strongly change the flows; e.g., a stepping-stone intruder can send traffic freely. They even can get root privilege on stepping stones and change the timing of packets. An intruder can test the presence of the marks before using

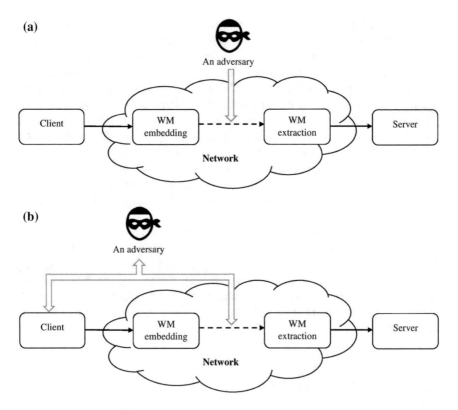

Fig. 11.3 Treat models for **a** isolate adversary **b** chosen flow adversary

his/her workstations at stepping stones, e.g., PNR attacker [22]. It is normally performed by setting some trial connections.

11.7.1 Invisibility with Isolated Adversaries

The capability of an adversary to recognize the flows with marks from non-watermarked flows is called as invisibility. An invisibility game is formally utilized by
an adversary. Take into account two network flows of $S0$ and $S1$. These sets of
flows are formed by the same distributions on the flows. The same network jitter is
influenced on sets $S0$ and $S1$, while the watermarker manipulates flows in $S1$ and
the flows in $S0$ remain unchanged. A random generator makes a random $i \in \{0, 1\}$
by using a fair coin flip. Two or more flows from Si are given to the adversary.
Considering the output $i0$, the adversary wins the game if $i = i'$ is obtained. The
watermark is known as invisible when the probability of wining the invisibility
game of an adversary becomes non-negligibly greater than $\frac{1}{2}$.

11.7.1.1 Output-Only Detection Encoder

In addition to flow timing, there is a little extra information that the isolated adversary can obtain from the watermark during the detection process. As the adversary can only obtain the possible outputs of the watermark encoder, it is called as output-only detection. These adversaries seem to be the weakest, but their abilities to detect the watermark are still under study. So far, entropy tests, distribution tests, and other primitive analysis methods have been widely studied in the area of isolated passive invisibility. Such tests are generally performed on individual flows.

11.7.1.2 Multi-flow Attacks

The invisibility of IBW schemes [5, 17] has been overlooked by MFA [5] specially when the same watermark is embedded into multiple flows. The potential power of a passive isolated attacker is revealed by MFA. If the frequency of packets is illustrated, multiple cleared or crowded intervals can be seen in the aggregated histogram which seldom occurs without watermarking. RAINBOW and SWIRL have been examined in the literature when MFA is imposed to the watermarking system.

11.7.2 Invisibility with Chosen Flow Adversaries

The aim of the chosen flow adversaries is to observe the arbitrary flows of network before and after embedding the watermark. In the following, two major strategies are explained.

11.7.2.1 Known Flow Attack

This attack is performed by pure observers. As the observers have access to the network flows before embedding the watermark, the invisibility game is modified. Moreover, the adversary can find the actual jitter of the network and the watermarker, i.e., the jitter-based watermarking schemes can be detected. As an instance, the invisibility of RAINBOW was studied by applying Kolmogorov–Smirnov (K-S) test on jitter vectors and IPDs in [7]. K-S test does not rely on a priori knowledge of the data distribution. As a result, it cannot efficiently estimate the power of adversaries against RAINBOW.

11.7.2.2 Chosen Flow Attack

This attack utilizes flows with specific timing patterns. Then, the adversaries examine the distortion possibility added by the watermarker. The ability of an adversary to interfere with flow generation is observed with the invisibility game. By using chosen flow attack and sending network packets with certain timing, it is possible to detect the watermark delays from network jitter. Unfortunately, network flow watermarking schemes cannot tolerate chosen flow attacks.

References

1. Wu, H.-C., and S.-H.S. Huang. 2010. Neural networks-based detection of stepping-stone intrusion. *Expert Systems with Applications* 37(2): 1431–1437.
2. Houmansadr, A. 2012. *Design, analysis, and implementation of effective network flow watermarking schemes*. University of Illinois at Urbana-Champaign.
3. Houmansadr, A., and N. Borisov. 2011. SWIRL: A scalable watermark to detect correlated network flows. In *NDSS*. 2011.
4. Houmansadr, A., N. Kiyavash, and N. Borisov. 2014. Non-blind watermarking of network flows. *IEEE/ACM Transactions on Networking (TON)* 22(4): 1232–1244.
5. Kiyavash, N., A. Houmansadr., and N. Borisov. 2008. Multi-flow attacks against network flow watermarking schemes. In *USENIX security symposium*.
6. Lin, Z., and N. Hopper. 2012. New attacks on timing-based network flow watermarks. In *Presented as part of the 21st usenix security symposium (USENIX Security 12)*.
7. Houmansadr, A., N. Kiyavash., and N. Borisov. 2009. RAINBOW: A robust and invisible non-blind watermark for network flows. In *NDSS*.
8. Bavier, A.C., et al. 2004. Operating systems support for planetary-scale network services. In *NSDI*.
9. Syverson, P., et al. 2001. Towards an analysis of onion routing security. In *Designing privacy enhancing technologies*. Berlin: Springer.
10. Chaum, D.L. 1981. Untraceable electronic mail, return addresses, and digital pseudonyms. *Communications of the ACM* 24(2): 84–90.
11. Boyan, J. 1997. *The anonymizer-protecting user privacy on the web*.
12. Dingledine, R., N. Mathewson., and P. Syverson. 2004. *Tor: The second-generation onion router*. DTIC Document.
13. Oh, H.O., et al. 2001. New echo embedding technique for robust and imperceptible audio watermarking. In *Acoustics, speech, and signal processing. IEEE international conference on proceedings of (ICASSP'01)*. IEEE.
14. Back, A., I. Goldberg., and A. Shostack. 2001. *Freedom systems 2.1 security issues and analysis. White paper, zero knowledge systems*. Inc., May, 2001.
15. Zhang, Y., and V. Paxson. 2000. Detecting stepping stones. In *USENIX security symposium*.
16. Kumar, R., and B. Gupta. 2016. Stepping stone detection techniques: Classification and state-of-the-art. In *Proceedings of the international conference on recent cognizance in wireless communication & image processing*. Berlin: Springer.
17. Wang, X., S. Chen., and S. Jajodia. 2007. Network flow watermarking attack on low-latency anonymous communication systems. In *Security and privacy. IEEE symposium on SP'07*. IEEE.
18. Pyun, Y.J., et al. 2007. Tracing traffic through intermediate hosts that repacketize flows. In *INFOCOM 2007. 26th IEEE international conference on computer communications*. IEEE.
19. Yu, W., et al. 2007. DSSS-based flow marking technique for invisible traceback. In *Security and privacy, 2007. IEEE symposium on SP'07*. IEEE.
20. Donoho, D.L., et al. 2002. Multiscale stepping-stone detection: Detecting pairs of jittered interactive streams by exploiting maximum tolerable delay. In *Recent advances in intrusion detection*. Berlin: Springer.

21. Adelsbach, A., S. Katzenbeisser., and H. Veith. 2003. Watermarking schemes provably secure against copy and ambiguity attacks. In *Proceedings of the 3rd ACM workshop on digital rights management*. ACM.
22. Peng, P., P. Ning., and D.S. Reeves. On the secrecy of timing-based active watermarking trace-back techniques. In *IEEE symposium on security and privacy, 2006*. IEEE.
23. Sencar, H.T., and N. Memon. 2007. Combatting ambiguity attacks via selective detection of embedded watermarks. *IEEE Transactions on Information Forensics and Security* 2(4): 664–682.

Chapter 12
Hardware IP Watermarking

12.1 Introduction

Recently, System-On-a-Chip (SOC) technology can easily design and implement a full system on a single chip (chipset). In order to consume time and cost for designing SOC, Intellectual Property blocks (IPs) or virtual components are used. This IP designs require significant amount of time and effort to be developed and verified. However, this IP designs can be illegally distributed, modified, and copied which must be protected to assure authorship proof.

According to VSI Alliance IP protection [1], three major mechanisms are available to secure IPs including deterrent, protection, and detection. In deterrent mechanism, some legal means such as trade secrets, copyrights, and patents are applied to protect the ownership without any physical protection. In protection mechanism, encryption and license agreements are used to protect IPs from being used by unauthorized parties. In detection mechanism, IP fingerprinting and IP watermarking technologies are used to trace and detect both illegal and legal usages of IPs. Although IP watermarking introduces some overhead in IP designs, direct interactions with the system provide a clear tracking mechanism to be presented in fort of a court if required. Therefore, IP watermarking technology is a potential solution to protect ownership and sensitive copyright information. IP watermarking technology embeds information into physical circuit during physical design level by configuring specific functions of Lookup Table (LUT) onto Field Programmable Gate Array (FPGA). Figure 12.1 illustrates an overview on the main issues in hardware IP watermarking which are explained in this chapter.

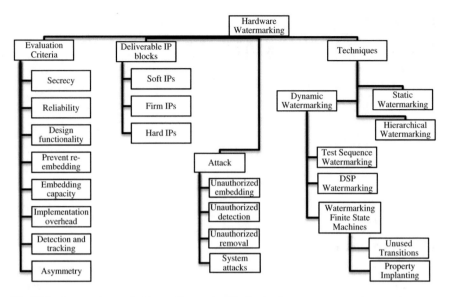

Fig. 12.1   An overview on hardware IP watermarking technology

12.2   Background of SOC Design Flow

The revolution in SOC designing can develop large and complex systems with wide functionalities. For reaching high amount of functionality and performance in SOC designing, both software and hardware must be integrated at early stage of design and development. Figure 12.2 presents the main levels from start to final stages in SOC design. Figure 12.2 is inferred from previous studies in [2, 3]. As shown, various high-level aspects such as system's requirements and system's specifications are determined at system level by considering both algorithmic and architectural designs. Algorithmic design is established by the main functionality of the end products. Afterward, architectural design is implemented algorithmic design by mapping and decomposing system specification into blocks by considering major components including operating system (OS), memories, and microprocessors. At lower level, partitioning modules assigns each function to specific software and hardware resources. As a result, the behavioral specifications of the system are modeled in behavioral level by applying designer programming languages at software part. Also, hierarchical hardware design flow is used to design IP blocks at various software/hardware design levels. As a conclusion, these IP blocks enable the engineers to reuse them and modify them with minimum time and cost.

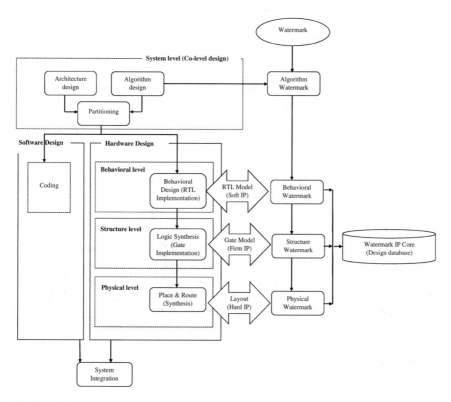

Fig. 12.2   IP watermarking in different levels of SOC design flow

12.3   Reusable and Deliverable IP Blocks

According to architecture document provided by Virtual Socket Interface (VSI), three major reusable and deliverable IP blocks are available including soft IP, firm IP, and hard IP (see Fig. 12.2). Choosing among them depends on some factors such as companies' commitment, applications, price, and cost. In the following, each of these deliverables is explained.

Soft IP: Soft IP is considered as the highest deliverable and reusable IP blocks which is in the form of synthesizable Hardware Design Language (HDL). Although soft IP is more flexible, performance of the system in terms of power, area, and timing cannot be estimated. Moreover, designing risk for soft IP is high due to integrator needs Register Transfer Level (RTL) source code.

Firm IP: Firm IP is located between soft IP and hard IP which provide a compromise among them. On the other word, firm IP can provide a deliverable IP blocks not only more flexible than hard IP, but also more predictable than soft IP in terms of area and performance. Thus, IP firm uses the combination of synthesizable RTL, a generic technology library, a full netlist, and detailed floorplan to optimize topology and structure of system in terms of area and performance. Lack of routing is the drawback of firm IP. The risk of designing for firm IP is less than soft IP if and only if RTL is not included.

Hard IP: Because high-level behavioral model is used minimally in hard IP, hard IP can optimize system in terms of mapping to specific technology, performance, size, routing, physical layout, and power. Moreover, hard IP is more predictable than soft IP and firm IP. Lack of using RTL code makes hard IP more protectable. However, hard IP is not portable and flexible due to dependencies among the processes.

12.4 Hardware IP Watermarking Techniques

According to literature in hardware IP watermarking, there are three main groups including dynamic watermarking, static watermarking, and hierarchical watermarking [2]. Static watermarking techniques consider a watermark as a property of the IP designs and extract the watermark by using static techniques such as placement and route watermarking. In dynamic watermarking techniques, the watermarked IP must be run to extract the watermark. Finite State Machine (FSM) and Digital Signal Processing (DSP) are applied to generate the watermark signal. Hierarchical watermarking techniques provide more secure system by using various design abstraction levels. In the following, different hardware IP watermarking techniques are discussed based on their advantages and disadvantages.

12.4.1 Dynamic IP Watermarking Technique

Dynamic IP watermarking techniques generate a signal during running the watermarked IP design to detect the watermark. In the following subsections, each dynamic IP watermarking technique is described.

12.4.1.1 Test Sequence Watermarking

Generally, this technique is effective when it is applied as complementary to other dynamic IP watermarking. In this technique, a random test sequence is searched for IP watermarking [4]. The watermark is generated automatically by integrating a watermark producer in a circuit on SOC. In this technique, the ownership can be easily verified without examining photomicrograph. Although this technique is novel, some issues should be noticed. Despite using test circuit instead of watermarking IP, the secrecy of the algorithm should be preserved. An adversary cannot remove the watermark by adding his/her watermark and the performance of the IP design is not degraded. As a result, the robustness is high and the overhead is low but asymmetric mode cannot be used. Furthermore, removing the watermark is direct and simple which makes calculation of masking and removal probabilities impossible.

12.4.1.2 Digital Signal Processing Watermarking

DSP watermarking technique is applied at the algorithmic level by changing decibel (dB) of the filter minimally [5, 6]. Thus, a digital filter is designed based on 7 partitions at high level to encode 7 bits of a watermark (character). On the other word, each part of the filter modulates a signal based on watermark bits. If the watermark bit is one, small change is done in dB of the signal. Otherwise, no dB change is needed for watermark bit is equal to zero. The lack of practical capacity for watermarking and difficulty for watermark tracking and detection are the main disadvantages of DSP IP watermarking technique.

12.4.1.3 Watermarking Finite State Machines

FSM watermarking technique can add new input and output sequences to IP design at behavioral level [7]. FSM IP watermarking technique can detect the watermark even from the lower level of IP design. Two FSM IP watermarking techniques include unused transitions and property implanting which are going to be discussed in the following parts.

(a) **Unused Transitions**

In this technique, unused transitions are extracted from State Transition Graph (STG), a behavioral level [7]. Then, new input and output sequences are inserted to STG for embedding the watermark. For this technique, the maximum number of free transitions (N_{max}) and minimum number of needed transitions (N_{min}) of unwatermarked design are computed. The probability of coincidence should not be satisfied for embedding the watermark which is adding input and output to STG. This approach does not need secrecy and can be detected even at lower levels of design. The robustness of this approach is high against FSM reduction due to variables used are becoming parts of other transitions. Sometimes, FPR is used to measure the robustness of this technique. An adversary can delete some transitions in order to degrade the watermark extraction process. Although Monte Carlo search and exhaustive search can be applied to decrease overhead of STG for an NP-hard problem, fining input sequence which can satisfy probability of coincidence is very difficult. In conclusion, high amount overhead and low embedding capacity are main problems for this technique.

(b) **Property Implanting**

In this technique, the watermark is embedded into FSM by implanting the STG implicitly [7]. For this purpose, an arbitrary and long watermark is encrypted by applying a one-way hashing function to an input sequence. Then, IP watermarking based on property implanting divides the input sequence into combination of sequences. For example, for 32 inputs and 256 bits sequence, 8 combinations of sequence are segmented. Next, these combinations of sequences are changing the STG states somehow a specific property is becoming rare in non-modified STG. A systematic way for changing these STG properties has been proposed [7]. Because it is not needed to find unused transitions in FSM, this technique has less overhead. Although IP watermarking based on property implanting can provide enough capacity for embedding, robustness (against masking attack and state reduction) and secrecy of this technique are low.

12.4.2 Static IP Watermarking Technique

Basically, static IP watermarking is based on constraint which can be embedded at various levels of the IP design [8]. This technique is the dominant IP watermarking approach that can be applied in each IP design level. For instance, linear programming problems [9], graph partitioning problems [10], and memory graph coloring problems [6] can be used at designing system level. Although some properties of this approach are dynamic, this approach is considered as a static approach. In order to constrain satisfaction (SAT) and optimize solution, hard NP problems are solved by available tools and adding extra constraints. The main components in static IP watermarking include:

(a) An NP-hard problem which must be optimized by adding some heuristics and constraints.
(b) An optimization algorithm for solving NP-hard problem
(c) Some constraints to be applied to IP design.
(d) Some extra constraints (well-formed grammars) to be added to prior IP design for producing watermarked IP design. For this purpose, a one-way encryption function maps the watermark to some well-formed constraints.

Two main efforts have been done for static IP watermarking based on localized watermarking and fair watermarking. Localized watermarking is dividing the watermark into small watermark components. Fair watermarking is embedding some parts of watermark to preserve the quality. Static technique can be extended for fingerprinting by dividing the IP design into two parts [11]. In the first part, before solving the problem, some relaxable constraints are introduced independently. The second part solves and optimizes these constraints for providing wide range of solutions. However, iterative optimization can also be applied for solving SAT problem. Adding watermark into nonlinear problem makes this approach robust and reliable. However, overhead due to iterations for watermarking, low quality of the watermarked design, and difficulty for detecting watermark are limitations of static approach.

12.4.3 Hierarchical IP Watermarking Technique

As mentioned, hierarchical IP watermarking technique embeds multiple water-marking in different design levels [12]. Because removing multiple watermarks in different design levels requires more effort, hierarchical IP watermarking is more robust and secure. However, this technique has high overhead on final design. Moreover, the designer should consider different issues during developing modules to assure about robustness, security, and minimum overhead.

12.5 Attack on IP Watermarking

IP watermarking is exposing to different attacks which can make significant high security risks for IP designs. There are four major groups of attacks such as system attacks, unauthorized detection, unauthorized embedding, and unauthorized removal that can threat security of IP watermarking [2]. In the following, each group of attacks is described.

System attacks: This attack is a serious attack which cannot be protected by watermarking technology. In this attack, the concept of watermarking is removed physically by an adversary. For instance, an adversary tries to destroy the chip which is verified by the watermark data. Thus, the watermark designer cannot do more to prevent this type of attack.

Unauthorized embedding: This attack tries to embed another watermark in IP designs by ghost searching technique. For this purpose, an adversary finds re-embeds a ghost watermark by applying necessary tools.

Unauthorized detection: This attack cannot seriously threat the security of IPs. This attack tries to extract the watermark data from IP design.

Unauthorized Removal: This attack tries to remove the watermark data without finding secret key during embedding time. Two main groups of unauthorized removal attacks include masking attacks and elimination attacks. In contrast to elimination attacks that attempt to remove the watermark completely, masking attacks try to distort the watermark extraction process. Similar to compression and denoising for detecting watermark from multimedia content, FSM reduction is applied to remove the watermark data from IP designs.

12.6 Criterion in IP Watermarking

Similar to multimedia watermarking criterion, IP watermarking can be evaluated depending on specific requirements of SOC and hardware designs. Some of these criterion evaluations are described in the following:

(a) *Secrecy of the algorithm.* The IP embedding and extraction algorithms should not depend on secrecy. It means any enemy can steal algorithms and the system should not put in trouble. The authorship data is protected by system properties.

(b) *Reliability.* Reliability is significant measurement and it is depending on robustness and FPR which are related to attack analysis. The hidden watermark should have enough strength against attacks as well as the watermark should not be detected from a non-watermarked design.

(c) *Design functionality.* The watermark should effect system behavior by proving its sound. On the other word, the watermark should be proved mathematically to have soundness.

(d) *Prevent re-embedding.* The watermarked IP designs should be robust against embedding another watermark which can degrade the authenticity of the original watermark in front of a court. Therefore, an adversary should not simply re-embed a watermark in the watermarked IP designs.

(e) *Embedding Capacity.* The size of watermark data should be enough large to be presentable in front of a court and should be enough small to keep system's overhead low. Thus, capacity is the main measurement that can be compared across different IP watermarking techniques.

(f) *Implementation overhead.* IP watermarking always increases the overhead of the system in terms of delay, power, and area. The amount of overhead is measured by comparing original design and watermarked IP design.

(g) *Detection and tracking.* Detecting and tracking the watermark is the critical process in IP watermarking. Although embedding watermark is half of the IP watermarking technology, the important part of any IP watermarking is extraction of the watermark data even under attacks. Thus, IP watermarking techniques are always judged in terms of detection and tracking measurement.

(h) *Asymmetry.* Asymmetric watermarking is a major challenge in IP watermarking as IP can be shared across third parties who can delete the watermark. Thus, any knowledge about watermark and key can caused to still and leakage of the watermarked IPs.

12.7 Discussion

In order to judge about pros and cons of each IP watermarking techniques, several properties are selected to compare among different IP watermarking techniques. Unlike image and audio watermarking that apply benchmark such as StirMark, there is no any benchmark for hardware IP watermarking. Therefore, the designers use FPR, FNR, overhead, cost of embedding, and cost of tracing to measure strength and robustness of the IP watermarking techniques. Table 12.1 presents an overall comparison among different IP watermarking techniques based on different criteria (some parts of Table 12.1 are from [2]). As inferred from Table 12.1, although static IP watermarking technique requires less embedding cost and low

Table 12.1 Comparison among different hardware IP watermarking techniques

Techniques		Security	FPR	Tracking cost	Overhead	Embedding cost
Static		H	L	H	L	L
Dynamic	DSP	H	H	H	L	L
	Property implanting	M	L	L	M	M
	Unused transitions	M	L	L	L	H
Hierarchical		H	L	H	H	H

H High
M Medium
L Low

overhead cost, it is suffering from a simple tracking approach. In addition, there is no possibility to add extra constrains for optimum solution. Both unused transitions and property implanting IP watermarking techniques can be detected and tracked the watermark during all cycles of IP block designs. However, unused transition technique has high embedding cost due to explosion in state space. Also, property implanting technique has problem with reduction techniques. On the other hand, DSP IP watermarking technique has less overhead and low embedding cost but it still suffers from high FPR due to low embedding capacity. Hierarchical IP watermarking technique not only can start from high levels of designs, but also it can integrate the watermarking through many design levels. Therefore, it seems hierarchical IP watermarking techniques would be the future for developing robust and secure final watermarked product.

References

1. Group, I.P.P.D.W. 2001. *Intellectual property protection: Schemes, alternatives and discussion*. VSI Alliance, White Paper, Version, 2001. **1**.
2. Abdel-Hamid, A.T., S. Tahar., and E.M. Aboulhamid. 2003. IP watermarking techniques: Survey and comparison. In *Proceedings of the 3rd IEEE international workshop on system-on-chip for real-time applications 2003*. IEEE.
3. Liang, W., et al. 2011. A chaotic IP watermarking in physical layout level based on FPGA. *Radioengineering* 20(1): 118–125.
4. Fan, Y.-C., and H.-W. Tsao. 2003. Watermarking for intellectual property protection. *Electronics Letters* 39(18): 1316–1318.
5. Chapman, R., and T.S. Durrani. 2000. IP protection of DSP algorithms for system on chip implementation. *IEEE Transactions on Signal Processing* 48(3): 854–861.
6. Hong, I., and M. Potkonjak. 1998. Techniques for intellectual property protection of DSP designs. In *Proceedings of the 1998 IEEE international conference on acoustics, speech and signal processing, 1998*. IEEE.
7. Oliveira, A.L. 2001. Techniques for the creation of digital watermarks in sequential circuit designs. *IEEE Transactions on Computer-Aided Design of Integrated Circuits and Systems* 20(9): 1101–1117.

8. Kahng, A.B., et al. 2001. Constraint-based watermarking techniques for design IP protection. *IEEE Transactions on Computer-Aided Design of Integrated Circuits and Systems* 20(10): 1236–1252.

9. Megerian, S., M. Drinic., and M. Potkonjak. 2002. Watermarking integer linear programming solutions. In *Proceedings of the 39th annual design automation conference*. ACM.

10. Wolfe, G., J.L. Wong, and M. Potkonjak. 2002. Watermarking graph partitioning solutions. *IEEE Transactions on Computer-Aided Design of Integrated Circuits and Systems* 21(10): 1196–1204.

11. Qu, G., and M. Potkonjak. 2000. Fingerprinting intellectual property using constraint-addition. In *Proceedings of the 37th annual design automation conference*. ACM.

12. Charben, E. 1998. Hierarchical watermarking in IC design. In *Proceedings of the IEEE custom integrated circuits conference*. Citeseer.

Chapter 13
Security Enhancement of Digital Watermarking

13.1 Introduction

This chapter discusses the security enhancement applications of digital water-marking in various technologies. Furthermore, the combination of watermarking technology with other computer security technologies such as biometric and cryptography is discussed. This chapter discusses the remaining issues in digital watermarking science.

In order to enhance the security of transmission channel, a cryptographic algorithm is applied on data. Cryptography and watermarking are conventionally applied on a layered architecture as shown in Fig. 13.1. As multiple lines of defense are provided by using this architecture, the system becomes more resilient to malicious attacks in comparison with a single line of defense.

13.2 Digital Watermarking Application Scenarios for Online Biometric Recognition Systems

Watermarking technology has improved the security of biometric systems [2, 3]. Watermark techniques have been studied in [4] by considering possible threats, requirements, and attacks. The summary of mentioned criteria for different watermarking techniques is shown in Table 13.1. In the following, each scenario is fully explained.

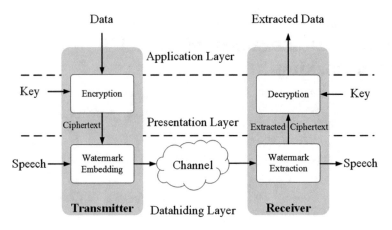

Fig. 13.1   Different layers in designing a watermark system [1]

Table 13.1   Performance criteria of watermarking in certain digital applications

Applications of WM	WM attack	Requirements of WM	Sample model(s)
Stenographic (covert) communication	Detection of the WM	Blindness, capacity, and fragility	[5]
Multi-biometric recognition	Distortion of modality of biometric data by the sample data	Blindness, capacity, and fragility	[6, 13, 14]
Multi-factor authentication	Embedding the sniffed password and token in the biometric features	Blindness, robustness (copyright protection), and fragility (tamper detection)	[8, 9, 10]
Sample replay prevention	Removing the WM	Blindness and robustness	[11]
Sample authentication	Correct acquisition of sensor data for mimicking the WM insertion	Fragility	[12, 15]

13.2.1 Stenographic (Covert) Communication

Stenographic (covert) communication hides a template of data by an arbitrary carrier. For example, DWT and FFT are applied on the cover speech signal for hiding the parameters of the LP model [5]. The sensitivity of this approach is very high against malicious attacks, but it has a high capacity.

13.2.2 Multi-biometric Recognition

Multi-biometric recognition benefits from watermarking. It utilizes two different biometric modalities to enhance the performance of biometric recognition. However, this approach is not close to real conditions for two reasons. The transmitter produces the biometric template (i.e., speaker feature). Therefore, the biometric features may be distributed in the systems with low sensor power. Moreover, the carrier of the biometric modality is subject to a high distortion injected by the biometric sample, during the embedding process.

Multi-biometric recognition has been utilized for embedding MFCC of the same individuals into a face image [6]. It was also used for embedding the voice feature into the iris image in order to improve the security by using watermarking and cryptography [7].

13.2.3 Multi-factor Authentication (MFA)

The performance of biometric recognition systems can be improved by the addition of multiple factors. These factors are normally extracted from knowledge-based or physical token methods. Although MFA is generally similar to multi-biometric recognition, they use different sensors and factors to enhance the security and recognition performance of the authentication method. As an instance, a two-factor authentication model was introduced in [8] that embedded a semi-fragile watermarking on iris image in smart cards.

The robustness of MFA against unintentional attacks can be improved by redundant embedding in the watermarks. However, addition of extra watermark bits can reduce the performance of recognition. For embedding the signature (as a password) into a face image in [9], high robustness of MFA was beneficial.

13.2.4 Sample Replay Prevention

Sample replay prevention points a scenario where the sensors are able to extract the watermark before the embedding. This approach keeps the biometric characters safe and protects the system against the fake inputs. An application of this model can be found in [11] where the process of speaker recognition is performed by verifying the watermark before feature extraction.

13.2.5 Sample Authentication

Sample authentication guarantees the geniality of the transmitter. In line with this, the integrity of the samples is checked by inserting the watermark to the biometric samples. A sample of this model [12] benefits from semi-fragile watermarking to verify the origin of the speech.

The above-mentioned watermarking applications are presented in Table 13.1, along with the behavior of attacks and the requirement of the watermarking and sample(s) of each application from the literature.

13.3 Biometric Watermarking

Nowadays, biometric recognition is becoming more popular. However, several breaches have been discovered which made biometric technology more vulnerable. Biometric watermarking tries to embed biometric templates as watermark in order to enable biometric recognition after the watermark detection. The summary of biometric watermarking features along with the major criteria for their comparison is presented in Table 13.2. It is shown that there is no blind watermarking technique with large value of capacity. Furthermore, almost all efforts have been performed in wavelet domain.

Signature and face image have been combined for recognition purposes as in multimodal approach [23]. Two cascading lifted wavelet transforms (LWTs) are applied for this purpose. The two biometric features can be offline handwritten signature and face that are inserted in a host image.

Developed image watermarking combines face image and fingerprint by inserting the face image in the fingerprint image [24]. Fingerprint image acts as the carrier of the facial features. This approach is a robust technique for improving the security of biometric system through application of DWT and LSB. The performance of this method is more than the sole DWT or LSB. However, it does not tolerate geometrical attacks.

A reverse form of developed image watermarking was introduced in [25], which utilizes the facial image as the carrier of fingerprint. For this purpose, block-wised

Table 13.2 Summary of biometric watermarking techniques

Ref.	Methods of WM	Types of WM	Blindness	Capacity
[16]	Spatial	Fingerprint + Eigen-face	Blind	Small
[17]	DCT	Iris binary code	Blind	Small
[18]	Correlation analysis	Palm	Non-blind	Large
[6]	RDWT	Voice	Semi-blind	Large
[19]	LSB + DWT	Iris	Blind	Small
[20]	DFT	Fingerprint	Blind	Small
[21]	DFT	Iris	Blind	Small
[22]	Wavelet	Face	Blind	Small

DCT and orthogonal locality preserving projections (OLPPs) have been applied for watermark embedding and watermark extraction, respectively. Note that OLPP gets a projection matrix of the watermark data.

Multimedia approach combines face, voice, and iris without watermarking [26]. This approach is of high time and cost complexity but improves the accuracy. A combination of voice and face was introduced in [27]. It performs a fusion detection approach that improves the verification performance.

Multimodal person authentication [28] combines facial features and MFCC. In order to embed MFCC into the face image, double-image watermarking by using DCT and QIM is performed. Double-image watermarking applies to both fragile and robust watermarking. The fragile watermark operates on salient facial regions including eyes, nose, and mouth areas, while the robust watermark covers the background of the face image. This approach seems efficient due to the combination of fragile and robust watermarking. However, it needs a huge capacity of watermark data that makes the technique infeasible.

A different multimodal approach has been presented in [6]. This approach combines voice and face features. The most important areas of the face are kept by phase congruency, and a three-level redundant discrete wavelet transform (RDWT) embeds mel-frequency cepstrum coefficient (MFCCs) features of the voice into the blue and red channels of the color face image.

Another multimodal approach, based on the double-image watermarking, utilizes a three-stage image watermarking [13] for inserting LPC of the voice, Gabor face, and Chaotic logistic map into an image host. This approach combines voice, facial, and offline handwritten signature. This approach hits a high rate of verification, but increases the complexity and number of required sensors. Furthermore, a three-stage watermarking corrupts the quality of the watermarked image seriously and makes the approach impractical.

The summary of multimodal approaches that utilized digital watermarking techniques is delineated in Table 13.3. The benefits and weaknesses of each approach are also presented in the table. Most of the approaches in Table 13.3 utilized face and voice features to implement the recognition system; however, a strong digital watermarking technique with acceptable robustness, imperceptibility, and capacity is not presented. In general, watermark embedding can decrease the performance of the face recognition system. Although the combination of voice and face features is common in multimodal recognition systems, it needs further efforts to improve the security of the system.

13.4   Quantum Watermarking

Quantum computing has benefits for watermark embedding [29]. Quantum watermarking is the area that embeds a quantum message M into the signal in the form of the watermarked message $\hat{M}$. A watermarked message $\hat{M}$ contains qubits. Considering I as the set of bits in M, the qubits $|\emptyset_i\rangle \in I$ are observed to watermark

Table 13.3 Summary of multimodal approaches by using digital watermarking techniques

Multimodal approach	Watermarking technique	Benefit	Weakness
Face + Fingerprint [24]	LSB + DWT	Simplicity	Low robustness against geometrical attacks, semi-blindness, and accuracy
Iris + Face + Voice [26]	None	High verification rate	No watermark, low security, and large complexity of time, cost, and memory
Face + Voice [27]	None	Simplicity	Low verification rate and security
Face + Handwrite Signature [23]	LWT	Simplicity	Double watermarking yields to high capacity of image watermarking, lack of integrated image watermarking technique
Face + Voice [28]	DCT-QIM	Good rate of verification, robustness, and tamper detection	Revoking the quality of host image after watermarking, low feasibility, and security Low security and feasibility
Face + Fingerprint [25]	Block-wise DCT	Low watermark data, and an efficient extraction process	Low rate of verification and image imperceptibility
Face + Voice + handwritten signature [13]	DCT-DWT	High rate of verification	Large complexity of time and memory, low cost, and low image imperceptibility
Face + Voice [6]	RDWT	High imperceptibility	High complexity, low robustness, and imperceptibility

M where i $\in$ I. All qubits $|\emptyset_i\rangle \in M$ are written in some basis j. In order to write qubits back to M, a dissimilar basis k is required. The result of this step is the watermarked message, $\hat{M}$. Therefore, $\hat{M}$ contains the original qubits in basis j and the watermarking qubits in basis k at $|\hat{\emptyset}_i\rangle \in \hat{M}$. Note that $k \neq j$ and $i \in I$. The watermark is produced when $\hat{M}\dot{}j$ ($\hat{M}$ is observed in the basis j). In this case, the intended value of error with the probability of error pe is the watermarking qubits written in basis k. By taking I and k as the secret keys, for $i \in I$, the relative frequency of error in the bits $a_i \in \hat{M}\dot{}j$ is the watermark.

Figure 13.2 illustrates an instance where qubits, basis states, and current state of a qubit are represented by circles, dashed lines, and solid lines, respectively. A is the initial state where the messages are encoded as qubits. The qubits in gray color (i.e., $I = 2, 3, 4, 6$) are rewritten in a secret basis by Alice. B is then sent to

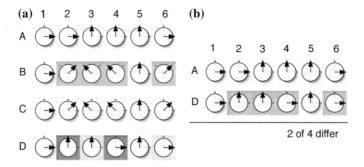

Fig. 13.2 Quantum watermarking **a** embedding process **b** watermark detection process [29]

Bob. Bob does not have any knowledge from the qubits that are written in a different basis. Then, he interprets the message in C. Bob obtains D when he observes the message. D is a fuzzy watermark with some errorful watermarking qubits, which is highlighted in dark gray, and some accurate watermarking qubits with light gray qubits. The watermark verification process is shown in Fig. 13.2.

Prior sending in verification process, M is watermarked. Therefore, the result of comparing $\hat{a}'_i \in \hat{M}' \cdot j$ with $a_i \in \hat{M} \cdot j$, where $i \in I$ is useful to detect whether $\hat{M}'$ is the watermarked of M. If the frequency of error between a_i and $\hat{a}'_i$ becomes close to the probability of errorful reading of a qubit, $\hat{M}'$ is probably the watermark of M by Alice. It is written in basis j when it was actually written in basis k. For the possibilities of the system, the term probably is used because the performance of the system is associated with the probabilities of qubits' observation. The following instance clarifies this sentence.

A qubit might have the value $|\emptyset\rangle = a_0|45°\rangle + a_1|135°\rangle$ in a certain basis. The observed bit might have an errorful value once $|\emptyset_i\rangle$ is read in a different basis from the original basis. Let us show this with $a_0|0°\rangle + a_1|90°\rangle$. $|\emptyset\rangle = \sqrt{0.5}|0°\rangle + \sqrt{0.5}|90°\rangle$ when $|\emptyset\rangle = a_0|45°\rangle + a_1|135°\rangle$. Thus, a value is read in the second basis with the probability of 0.5 from the expected read value in the first basis. The relative frequency of error is expected to be near to the expected probability of error, on the condition that enough qubits written in one basis are read in another basis. A number of n bits in minimum are needed when $0 \leq pe \leq 1$ and $n = a + b$ s.t. $\frac{a}{a+b} \geq pe$ bits.

Figure 13.2 represents the way that Alice realizes D is the same message A on which she had embedded the watermark. For this purpose, she compares two messages on the qubits in I and computes the relative error frequency. When 2 out of 4 bits are erroneous, a relative frequency of 0.5 is obtained. If $pe = 0.5$ is taken into account from Fig. 13.2, then Alice realizes D as the watermarked A.

Normally, more than one watermark, i.e., $\hat{M}$, is assumed for a single message M. In this case, an adversary needs to calculate the average of multiple watermarks to estimate the original message before watermarking. The adversary can do this because only one difference is available for each $\hat{a}'_i \in \hat{M}' \cdot j$ (where $i \in I$)

from a_i with the probability of pe. The more the versions from $\hat{M}$ for an adversary, the more the possibility to find the original message is. He observes the positions where the bits are different and guesses the value of k. By taking the average values of watermarking bits, a version can be created by the adversary with a different imperceptibility from the original version without a valid watermark. When for a message M, various values of I is available for each $\hat{M}$, finding the average value gets harder, not impossible. This scheme is based on the frequency value where a_i disagrees with $\hat{a}_i'$. Therefore, for small values of $|I|$, the number of unchanged $\hat{a}_i' \in \hat{M}' \cdot j$ is decreased for an enough and imperceptible amount of noise. It is on the condition that the relative frequency of error does not get a value near the expected probability of error. Moreover, the relative frequency of error may not get close to pe when values of $\hat{M}_i'$ are modified. If the value of $|I|$ becomes larger and less than the perceptibility of the watermark, the scheme can tolerate the attacks that make the message useless.

Some attacks change the indices of bits with no message rendering. It is performed by padding the message with non-destructive bits, e.g., the bits of a message, $\tilde{a}_1, \tilde{a}_2, \ldots, \tilde{a}_n$ for $\tilde{a}_i \in \hat{M}$, which can be shifted by $+1$; thus, $\tilde{a}_i$ will be $\tilde{a}_i + 1$ and $\tilde{a}_0$ becomes imperceptible or ignored. For detecting this attack, the transform should be discovered and reversed. Thus, a_i should be matched with $\hat{a}_i'$.

13.5 DNA Watermarking

Nowadays, bioinformatics science has applied computer algorithms to process biological data. In this line, Deoxy-ribonucleic Acid (DNA) can be emerged as a carrier for hiding watermark information. Basically, DNA consists of quaternary finite strings including Guanine (G), Thymine (T), Cytosine (C), and Adenine (A) that are bonding between two opposite oriented nucleotides [30]. DNA watermarking can be used to detect any unauthorized usage of genetically modified organisms (GMOs). Also, it can be used for covert communications. However, DNA has certain biological constraints that cannot be modified arbitrary. For instance, two pairing of C-G and A-T cannot change. Moreover, mutations in the watermarking must be considered to prevent any permanent modification in the structure of a DNA. As a result, the watermark data is embedded into non-codded parts of the DNA. DNA-Crypt program is corrected predicted mutations by applying WDH-code and Hamming-code in order to preserve encrypted information intact. Available DNA watermarking techniques have been synthetized DNA string to embed the binary watermark data. Although these synthetic string can be applied for authentication purposes, DNA string can be influenced on living organisms. In addition, mutations are degraded the watermarked DNA due to its infrequent manner. Some of the DNA watermarking techniques have been embedded the watermark data based on LSB replacement method. Another DNA watermarking technique has been proposed based on DWT to transform DNA string in frequency domain. Then, histogram ranks method has been used to code some DNA subsequences [31]. Moreover, DNA watermarking technique can be combined with cryptographic algorithms such as Blowfish, RSA, and AES to improve the security of DNA. There are some requirements for design a DNA watermarking technique including: non-blind watermark detection (because the knowledge of reference genome strings is essential for watermark data detection), codon optimization (DNA watermarking must be considered inverted/direct repeats, GC content, DNA motifs, codon text, and codon usage for gene analysis and gene expression), security, amino acid conservation, transparency, capacity, and mutation resistance.

13.6 Application of Watermarking in Financial Systems

Due to high sensitivity and low cost security solution, watermarking can be effectively applied for financial systems. However, a few research studies have been conducted for this purpose. For instance, the ownership of Bitcoin, as decentralized electronic currency system, requires an effective mechanism to be preserved. Thus, new approaches have been developed to watermark Bitcoin which is known as colored coins [32, 33]. This watermarking approach can be extended for assets such as stock, car, and smartphone. In addition, Bitcoin blockchain can also be used to track them. Digital watermarking can be used as antimony-laundering technique by tracing suspect transactions and possibly determining the degree of separation between suspects [34].

13.7 Robustness Versus Security

Digital watermarking techniques are mostly seen from two different aspects: security and robustness. For a long time, there was no effort to distinct the properties of watermarking techniques regarding these two concepts. Security targets the performance of the watermarking technique in a noisy or hostile environment. On the other hand, the effect of routine parameters of signal processing is concentrated on the watermarked multimedia.

Lossy compression as widely used in multimedia transmission and data storage eliminates certain parts of multimedia and increases the probability of error in data retrieving or missing the watermarked content. However, it is not accounted as a security threat when regular customers interact with the system. In a nutshell, security is an issue when regular customers are not the sole users of a system and the system is subject to adversaries' attacks. While robustness concerns the satisfaction of regular users from the performance of the system, robustness attacks are particularly universal or generic. The concepts of robustness and security can be discussed based on the intention of users; e.g., a hacker or intruder attempts to obtain particular information from the watermarked data of a whole database

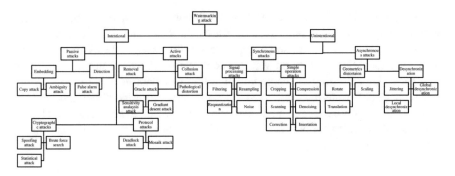

Fig. 13.3 Classification of watermarking attacks according to robustness and security

system while a customer blindly compresses a JPEG image and may unintention-
ally remove the watermark. Moreover, the collected information by an attacker is
later misused to remove, detect, or edit the embedded watermarks and to design
dedicated attacks which are not limited to watermark removal. In summary,
robustness attacks concern about the watermark removal, while security attacks
also concentrate on unauthorized actions on watermark, e.g., modification, writing, estimation, and detection.

Classification of digital watermarking attacks according to robustness and security is brought in Fig. 13.3. It is an extended form of a previous classification in
[35, 36]. Robustness attacks are further classified into synchronous and asynchronous attacks. Synchronous attacks cover common signal processing primitives that
influence the watermark signal and delay the detector. They are lossy compression,
filtering, denoising, and quantization. Asynchronous attacks modify the positions
of samples but do not explicitly remove the watermark. Therefore, the decoder
becomes useless, and the synchronization convention is shared by the embedder
and detector. As the watermark cannot be retrieved by the detector, the asynchronous attacks are counted as removal attacks [37, 38].

Random bending attack (RBA) [35] is a famous reference attack to measure
the robustness of a watermarking technique. For evaluating the effect of ADC
and DAC convertors, RBA applies to geometrical distortions. Sensor alignment
and lens parameters are changed for the application of geometrical distortions.
Moreover, the response of non-ideal sensors and JPEG compression is simulated
by the addition of noises.

Security attacks of digital watermarking techniques are classified into protocol and cryptographic attacks. Protocol attacks discover and misuse the general
knowledge about the frame of watermarking. As an instance, a copyright attack
can be found when the multimedia item carries more than a single watermark.
However, it is not always possible to detect which watermark was embedded originally [39] and who is the owner of the multimedia. To detect the fake ownership
of the documents with copyright, search robots (Web crawlers) that automatically
search the Web for illegally copyrighted documents can be utilized. In order to

mislead the Web crawlers, one attacker can split the images into small pieces and insert them in a proper sequence in a Web page [40]. This strategy interprets the original image as a stuck of images which confuses the watermark detector.

Discovery of the secret key of the watermark signal (or the pseudorandom sequence) is the main goal of cryptographic attacks. Brute-force search is an attack in this category which tests all the potential keys to discover the correct key. Another instance is steganalysis which enables unauthorized detection of watermark by isolating some characteristics of the watermarking method [41]. Oracle attacks and statistical attacks (or collusion attacks) are other samples of cryptographic attacks. The watermarks are altered periodically by the Oracle attacks to make the publicly available detectors unable to recover the original watermark [42] while the watermarked documents are collected and combined together by statistical attacks in order to form a non-watermarked document. The copy attack is the last attack in this category by which a watermark document is selected, and then its watermark is estimated and is recovered to a non-protected document [43]. It is also counted as an unauthorized watermark writing.

References

1. Donoho, D.L., et al. 2002. Multiscale stepping-stone detection: Detecting pairs of jittered interactive streams by exploiting maximum tolerable delay. In *Recent advances in intrusion detection*. Springer.
2. Faundez-Zanuy, M., M. Hagmüller, and G. Kubin. 2006. Speaker verification security improvement by means of speech watermarking. *Speech Communication* 48(12): 1608–1619.
3. Faundez-Zanuy, M., M. Hagmüller, and G. Kubin. 2007. Speaker identification security improvement by means of speech watermarking. *Pattern Recognition* 40(11): 3027–3034.
4. Hämmerle-Uhl, J., K. Raab, and A. Uhl. 2011. Watermarking as a means to enhance biometric systems: A critical survey. In *Information hiding*. Springer.
5. Rekik, S., et al. 2012. Speech steganography using wavelet and Fourier transforms. *EURASIP Journal on Audio, Speech, and Music Processing* 2012(1): 1–14.
6. Vatsa, M., R. Singh, and A. Noore. 2009. Feature based RDWT watermarking for multimodal biometric system. *Image and Vision Computing* 27(3): 293–304.
7. Bartlow, N., et al. 2007. Protecting iris images through asymmetric digital watermarking. In *2007 IEEE Workshop on automatic identification advanced technologies*. IEEE.
8. Huber, R., H. Stögner, and A. Uhl. 2011. Two-factor biometric recognition with integrated tamper-protection watermarking. In *Communications and multimedia security*. Springer.
9. Satonaka, T. 2002. Biometric watermark authentication with multiple verification rule. In *Proceedings of the 2002 12th IEEE workshop on neural networks for signal processing, 2002*. IEEE.
10. Nematollahi, M.A., H. Gamboa-Rosales, F.J. Martinez-Ruiz, I. Jose, S.A.R. Al-Haddad, and M. Esmaeilpour. 2016. Multi-factor authentication model based on multipurpose speech watermarking and online speaker recognition. *Multimedia Tools and Applications*, 1–31.
11. Lien, N.T.H. 2009. *Echo hiding using exponential time-spread Echo Kernel and its applications to audio digital watermarking and speaker recognition*. Tokyo Institute of Technology.
12. Yan, B., and Y.-J. Guo. 2013. Speech authentication by semi-fragile speech watermarking utilizing analysis by synthesis and spectral distortion optimization. *Multimedia tools and applications* 67(2): 383–405.

13. Inamdar, V.S., and P.P. Rege. 2014. Dual watermarking technique with multiple biometric watermarks. *Sadhana* 39(1): 3–26.
14. Vielhauer, C., et al. 2006. Multimodal speaker authentication–evaluation of recognition performance of watermarked references. In *Proceedings of the 2nd workshop on multimodal user authentication (MMUA), Toulouse, France.*
15. Zhe-Ming, L., Y. Bin, and S. Sheng-He. 2005. Watermarking combined with CELP speech coding for authentication. *IEICE Transactions on Information and systems* 88(2): 330–334.
16. Ratha, N.K., J.H. Connell, and R.M. Bolle. 2000. Secure data hiding in wavelet compressed fingerprint images. In *Proceedings of the 2000 ACM workshops on multimedia.* ACM.
17. Vatsa, M., et al. 2004. Comparing robustness of watermarking algorithms on biometrics data. In *Proceedings of the workshop on biometric challenges from theory to practice-ICPR workshop.*
18. Qi, M., et al. 2010. A novel image hiding approach based on correlation analysis for secure multimodal biometrics. *Journal of Network and Computer Applications* 33(3): 247–257.
19. Fouad, M., A. El Saddik, and E. Petriu. 2010. Combining dwt and lsb watermarking to secure revocable iris templates. In *2010 10th International conference on information sciences signal processing and their applications (ISSPA).* IEEE.
20. Khan, M.K., L. Xie, and J. Zhang. 2007. Robust hiding of fingerprint-biometric data into audio signals. In *Advances in biometrics*, 702–712. Springer.
21. Park, K.R., et al. 2007. A study on iris feature watermarking on face data. In *Adaptive and natural computing algorithms*, 415–423. Springer.
22. Noore, A., et al. 2007. Enhancing security of fingerprints through contextual biometric watermarking. *Forensic Science International* 169(2): 188–194.
23. Arya, M.S., and R. Siddavatam. 2013. Geometric robust multimodal biometric watermarking scheme for copyright protection of digital images. *International Journal of Computer Applications* 72(9): 40–52.
24. Vatsa, M., et al. 2006. Robust biometric image watermarking for fingerprint and face template protection. *IEICE Electronics Express* 3(2): 23–28.
25. Ghouzali, S. 2015. Watermarking based multi-biometric fusion approach. In *Codes, cryptology, and information security*, 342–351. Springer.
26. Sheetal Chaudhary, R.N. 2015. A new multimodal biometric recognition system integrating Iris, face and voice. *International Journal of Advanced Research in Computer Science and Software Engineering.* 5(4): 145–150.
27. Soltane, M., N. Doghmane, and N. Guersi. 2010. Face and speech based multi-modal biometric authentication. *International Journal of Advanced Science and Technology* 21(6): 41–56.
28. Wang, S., et al. 2013. Augmenting remote multimodal person verification by embedding voice characteristics into face images. In *2013 IEEE International conference on multimedia and expo workshops (ICMEW).* IEEE.
29. Gordon, G. and W. Iii. 2008. *Quantum watermarking by frequency of error when observing qubits in dissimilar bases.*
30. Panah, A.S., R. van Schyndel, T. Sellis, and E. Bertino. 2016. *On the properties of non-media digital watermarking: a review of state of the art techniques.*
31. Lee, S.H. 2014. DWT based coding DNA watermarking for DNA copyright protection. *Information Sciences* 273: 263–286.
32. Nakamoto, S. 2008. *Bitcoin: a Peer-to-Peer Electronic Cash System.* [Online]. http://www.cryptovest.co.uk/resources/Bitcoin%20paper%20Original.pdf.
33. Rosenfeld, M. 2012. *Overview of colored coins, white paper.* [Online]. https://bitcoil.co.il.
34. Leigh, D. 2016. *A test bed for data hiding in financial transactions.* [Online]. http://www.dylanleigh.net/portfolio/s3017239-summer-summary.pdf.
35. Petitcolas, F.A., R.J. Anderson, and M.G. Kuhn. 1998. *Attacks on copyright marking systems.* In *Information hiding.* Springer.
36. Voloshynovskiy, S., et al. 2001. Attack modelling: Towards a second generation watermarking benchmark. *Signal Processing* 81(6): 1177–1214.

37. Kalker, T. 2001. Considerations on watermarking security. In *2001 IEEE Fourth Workshop on Multimedia Signal Processing*. IEEE.
38. Petitcolas, F.A. (2004). *Stirmark benchmark 4.0*.
39. Craver, S., et al. 1998. Resolving rightful ownerships with invisible watermarking techniques: limitations, attacks, and implications. *IEEE Journal on Selected Areas in Communications* 16(4): 573–586.
40. Obaid, A.H. 2015. Information hiding techniques for steganography and digital watermarking. *UDC 681.518 (04) INTERACTIVE S < STEMS: Problems of human-computer interaction*. Collection of scientific papers, 306 p, 63. Ulyanovsk: USTU.
41. Chandramouli, R., M. Kharrazi, and N. Memon. 2003. Image steganography and steganalysis: Concepts and practice. In *Digital watermarking*, 35–49. Springer.
42. Linnartz, J.-P.M. and M. Van Dijk. 1998. Analysis of the sensitivity attack against electronic watermarks in images. In *Information hiding*. Springer.
43. Holliman, M., and N. Memon. 2000. Counterfeiting attacks on oblivious block-wise independent invisible watermarking schemes. *IEEE Transactions on Image Processing* 9(3): 432–441.

Printed in the United States
By Bookmasters